SP Pvt. Ltd.

*Dedicated
with love and special
regards
to our parents*

Tissue Culture Based *Sugarcane Farming*

R.S. Sengar
Department of Biotechnology, College of Agriculture
Sardar Vallabhbhai Patel Universtiy of
Agriculture and Technology
Meerut – 250 110

M. Lal
Sugarcane Research Institute
U.P. Council of Sugarcane Research
Shahjahanpur – 242 001

Shalini Gupta
Department of Molecular Biology and Genetic Engineering
College of Biotechnology
Sardar Vallabhbhai Patel University of
Agriculture and Technology
Meerut – 250 110

2013

Studium Press (India) Pvt. Ltd.

Tissue Culture Based *Sugarcane Farming*

ISBN: 978-93-80012-35-3

Published by:

Studium Press (India) Pvt. Ltd.
4735/22, 2nd Floor, Prakash Deep Building
(Near Delhi Medical Association),
Ansari Road, Darya Ganj, New Delhi-110 002
Tel.: 91-11-43240200-15 (15 lines), Fax: 91-11-43240215;
E-mail: studiumpress@gmail.com

Printed at : Akash Press, New Delhi

Foreword

Tissue culture is a very important and emerging area of biotechnology in 21st century. In the plant tissue culture the cells or tissues are obtained from any meristematic part of the plant and are encouraged to regenerate under controlled environmental conditions on artificial nutrient media. In recent years, there has been a large increase in the number of research laboratories using tissue culture techniques to investigate many fundamental and applied aspects of higher plants. The tissue culture techniques are now being exploited by many commercial laboratories. Even many horticultural companies are setting up small units to multiply plants which are difficult to propagate by conventional means.

Micropropagation is one of the most important and perhaps the most utilized technique of plant tissue culture. It involves establishment of axenic shoot cultures and production of large number of plantlets of a variety using specific plant parts within a limited period of time. In addition to successful application of *in vitro* micropropagation technique in many fruits, trees, aromatic and ornamental plants, it is now being used to multiply newly released varieties of sugarcane as well. The potential of tissue culture is amply recognized by its inclusions in some form or other in various course curricula. Therefore the present book entitled **"Tissue Culture Based Sugarcane Farming"** authored by **R.S. Sengar, Madan Lal** and **Shalini Gupta** appears to be very useful to the students of undergraduate and post-graduates. I hope the students will find this book useful as it provides comprehensive course material.

I appreciate the hard work put in by the authors in preparing this book. I congratulate them for their concern efforts and successful intellectual journey.

(Dr. R.S. Kureel)

Vice Chancellor
N.D. University of Agriculture & Technology,
Kumarganj, Faizabad, (U.P.)

Preface

Sugarcane (*Saccharum species* complex) is an important agricultural cash crop in tropical and subtropical region of the world. Being a major source of sugar it occupies a distinct position as an agro-industrial crop of the country, covering around 4.36 million hectares area (3% of the total cultivated area) with an annual cane production of about 300 million tonnes, contributing about 7.5% to the gross value of the agricultural production in the country. About 50 million farmers and equal number of agricultural laborers depend on sugarcane for their livelihood. Half a million skilled workers are also engaged in the sugar industry. A total of 453 sugar factories crushed 194.4 million tonnes of cane (69% of total cane production) producing 20.14 million tonnes of sugar. Average yield of sugarcane in India is about 70 tonnes per hectare. India ranks second in world sugar production next to Brazil. India is known as the original home of sugar and sugarcane. In global sugar economy, the Indian sugar industry has achieved a number of milestones. It is the second largest producer of sugar in the world. Sugar industry in India is well developed with a consumer base of more than billions of people. This industry is the second largest agro processing industry in India next to cotton and textile industry. Sugarcane occupies about 2.5 per cent of country's gross cropped area and provides raw material to more than 453 sugar mills for production of sugar and by- products. India is the largest consumer and second largest producer of sugar in the world next to Brazil, having a share of over 14 percent of the world sugar production. But in terms of yield, India stood at 5th place among major producers in the world. More than 500 thousand peoples are directly employed in this sector. Including farmers and their family members, more than 45 million peoples of the rural population in India depend on sugar industry for their livelihood. Its contribution to the central and state exchequers is of high order. The Indian sugar industry has been accounting for around 1% GDP of the country in the recent past. Presently about 5 million hectares of land is under sugarcane with an average yield of 70 tons per hectare. About 45 million Indian farmers and their families are engaged in the cultivation of sugarcane. In addition to it a large mass of agricultural labour is involved in sugarcane anciallary activities constituting 75% of the rural population. Sugar production in India is concentrated in six states namely Maharashtra, Uttar Pradesh, Gujarat, Tamil Nadu, Karnataka, and Andhra Pradesh, which together account for 85-90% of sugar production in the country. The Indian sugar industry is highly fragmented with over 450 mills and no single player having a market share of over 5%. Around 6% of the mills are in the cooperative sector, 35% in the private and the rest in public sector. The sugar industry can be broadly classified into the two sub-sectors, the organized sector *i.e.* sugar factories, and the unorganized sector *i.e.* manufacturers to the traditional sweetner like gur and khandsari. Sugarcane being propagated vegetatively has a low seed multiplication ratio (1:6 to 1:8) due to which seed production of newly released varieties is invariably slow. Further the seed perpetuates diseases and pests during several cycles of field production. Hence non-availability of disease - free, true to the type planting material is a major constraint in improving sugarcane productivity. Development of tissue culture technology for rapid multiplication of disease- free planting material has been an important step towards quality seed production in sugarcane. India, Australia and Philippines have already applied this technology for commercial seed production and the benefits have become evident through rapid multiplication and distribution of elite varieties and increased sugarcane production. The experiences of these countries would be of considerable benefit to all those who are in the process of adopting micropropogation for their seed production programs. The present publication provides detailed information on micropropogation and related field propagation techniques. It is hoped that this publication will be of immense use to the seed industry, researchers and policy planners in evolving efficient seed production system for rapid and wide coverage under improved varieties of sugarcane.

R.S. Sengar

Acknowledgements

I am deeply indebted to Prof. Lalji Singh Vice Chancellor BHU, Prof. D.P. Singh Vice-Chancellor, Prof. Shivendra Vikram Shahi, Western Ktencky University Bowling Green, USA, Prof. V.P. Singh, Dean, Dept. of Plant Science, M.J.P. Rohilkhand University, Bareilly and Late Prof. H.S. Srivastava, M.J.P. Rohilkhand University, Bareilly for their encouragement and desire to publish this experimental book for the use of students, scholars, teachers and researchers working in Biotechnology and related fields.

I wish to thank my department colleagues, students and friends for their help and support to complete this book. I extend special thanks and gratefully acknowledge the information in the form of figures, tables and others which have been adapted in the book from various sources.

I express my gratitude to all who have helped us directly or indirectly in writing this book. I also extend sincere thanks to my parents who always inspired me with the words "Try to leave behind your footprints on the sand of time through some of your humble contributions to the society".

I am also thankful to respected senior teachers who have imparted their knowledge on the subject. This would not have been a successful effort without their blissful inspiration and guidance.

I am indebted to the co-authors, Dr. M.Lal and Dr. Shalini Gupta for their help and co-operation in completing the manuscript.

I also thank the editorial staff members of Studium Press (India) Pvt. Ltd., Delhi for their active cooperation. I express my sincere gratitude to our friends and near and dear ones for their unfailing emotional support during the process of writing this book.

Finally, this book would never have been accomplished without the encouragement and inspiration of my mother Smt. Kamla Senger, Father Dr. S.S. Sengar, Brother Dr. Rajesh Singh Sengar, wife Sarita Sengar and great pleasure obtained from my two sons Divyanshu Sengar and Kartikey Sengar while preparing the book. I am also thankful to Miss Reshu Chaudhary, Miss Kalpana Sengar, Vevakanand Pratap Rao and other faculty members of the University who have helped in various ways.

All suggestion for further improvement of the book can be sent either at rs.svbpatu@gmail.com or sengar65@gmail.com

R.S. Sengar

About the Authors

Dr. R.S. Sengar is working as an Associate Professor officer Incharge of Biotechnology in Deptartment of Agriculture Biotechnology, College of Agriculture in Sardar Vallabhbhai Patel University of Agriculture and Technology, Meerut. Dr. Sengar has 18 years teaching, research and extension experience to his credit. He was associated with teaching for a period of about 10 years at G.B. Pant University of Agri. and Tech. Pantnagar (Uttrakhand). He has supervised 4 M.Sc. and Ph.D. students and published more than 45 research papers in India and Foreign Journals of repute. Dr. Sengar contributed 105 abstracts/papers to different conferences/symposiums/workshop organized at National and International forums. He has published 810 popular articles in Hindi and English languages in important journals and magazines. He has also published 6 books for scientific community and farmers. Dr. Sengar has delivered more than 45 talks at All India. Radio and Television at Delhi, Lucknow, Rampur and Bareilly centers. Dr. Sengar is life member of many professional societies. He is a member of the editorial advisory boards and review committees of few journals. He has been the Principal and Co–Investigator of various projects financed by different government departments. He has made several contributions on *in vitro* culture of sugarcane. Crop improvement through somaclonal variation and micropropagation of newly released varieties of sugarcane for rapid seed multiplication, are the fields of his main interest. Dr. Sengar has been the recipient of the 'Best Writer' award from Vishwa Agro Marketing and Communication Kota Rajasthan, 'Kunwar Saxena Bahadur SRDA' award from society for Recent Development in Agriculture Meerut, 'Man of the Year' 'Research Board of Advisors' awards from American Biographical Institute, Inc. U.S.A., 'Aryabhat' award from Vigyan

Bharti, New Delhi, Fellow of Society of Plant Research Bareilly, (U.P.), Gold Medal HTHS award from Hi-Tech Horticultural society Meerut and J.C. Edward Medal from BIOVED as Research Society Allahabad, Uttar Pradesh. He participated at International Conferences/Workshop/ Seminars as delegate organizer and resource person.

Dr. Madan Lal is presently working as a Scientific Officer in Tissue Culture Division. U.P.C.S.R., Shahjahanpur, India. After post graduation in Botany in 1982, he worked for his Ph.D with Professor (Late) V.S. Jaiswal, Banaras Hindu University, Varanasi. He has made several contributions on *in vitro* culture of sugarcane. He has emphasized on sugarcane improvement through somaclonal variation and rapid multiplication of newly released varieties of sugarcane through *in vitro* micropropagated. He has also worked on some biochemical and molecular changes taking place during *in vitro* morphogenesis in sugarcane. He has developed low cost protocols for efficient micropropogation of several varieties of sugarcane for seed purposes. He has conducted numerous field demonstrations, using micropropagated plantlets of sugarcane, at the farms of cane growers and sugar factories of U.P. and motivated them to adopt the micropropagated seed material. He has successfully encapsulated the somatic embroys and young meristems of sugarcane varieties and obtained germination of artificial seeds *in vitro*. Dr. Lal has published numerous research papers in Indian and International journals.

Dr. Shalini Gupta, M.Sc., Ph.D. from G.B. Pant Agriculture University, Pantnagar is presently working as Assistant Professor of Molecular Biology and Genetic Engineering, College of Biotechnology, Sardar Vallabhbhai Patel University of Agriculture and Technology, Meerut. She has a Teaching and Research experience of nearly 8 years. She is a good researcher and published several scientific papers/ articles and research papers in different National and International Journals of repute. She was honoured with the Young Fellow Award by the Society of Recent Development in Agriculture. Her areas of interest are Genetics, Molecular Biology and Molecular Breeding.

Contents

Abbreviations and Acronyms

APAARI	Asia Pacific Association of Agricultural Research Institutions
APCoAB	Asia–Pacific Consortium on Agricultural Biotechnology
AST	Aerated Steam Therapy
BAP	6–benzylaminopurine
BSEs	Bureau of Sugar Experiment Station, Brisbane, Australia
CCS	Commercial Cane Sugar
CD	Critical Difference
CMRTD	Consortium on Micropropagation Research and Technology Development
CoC	Coimbatore–Cuddalore
CoH	Coimbatore–Haryana
CoS	Coimbatore–Shahjahanpur
CoSe	Coimbatore–Seorahi
CoSi	Coimbatore–Sirugamani
CoJ	Coimbatore–Jalandhar
DBT	Department of Biotechnology, Government of India
DAC–ELISA	Direct Antigen Coating–ELISA
DAS–ELISA	Double Antigen Sandwich–ELISA
EDTA	Ethylenediaminetetraacetic acid
ELISA	Enzyme–linked Immunosorbant Assay
FAO	Food and Agricultural Organization of the United Nations
FYM	Farm Yard Manure

Ha	Hectare
HEPA	High Efficiency Particulate Air
IBA	Indole–3–butyric acid
ICAR	Indian Council of Agricultural Research
LDM	Liquid Differentiating Medium
MC	Meristem Tip Culture
MS	Murashige and Skoog
MCB	Methyl Benzimidazole–2yl–carbamate
mg/l	Milligrams/litre
mt	Million tonnes
NAA	*a*–Napthaleneacetic Acid
NPK	Nitrogen–Phosphorus–Potash
PCR	Polymerase Chain Reaction
pH	Hydrogen Ion Concentration
psi	Pounds per square inch
R&D	Research and Development
RT–PCR	Reverse Transcription Polymerase Chain Reaction
SARP	Sugarcane Adaptive Research Programme
SBI	Sugarcane Breeding Institute
SE	Standard Error
SCMV	Sugarcane Mosaic Virus
SCYLV	Sugarcane Yellow Leaf Virus

Chapter 1

Laboratory Organization for Plant Tissue Culture

Plant tissue culture is not a separate branch of plant science like taxonomy, cytology, plant physiology etc. Rather it is a collection of experimental methods of growing large number of isolated cells or tissues under sterile and controlled environmental conditions. The cells or tissues are obtained from any part of the plant like stem, root, leaf etc which are encouraged to produce more cells in culture and to express their totipotency (*i.e.*, their genetic ability to produce more plants). Cells or tissues are grown in different types of glass vessels containing a medium with mineral nutrients, vitamins and phytohormones. Therefore, to carry out the experiments using tissue culture techniques, a well–equipped laboratory is the first requirement.

In recent years there has been a large increase in the number of research laboratories using tissue culture techniques to investigate many fundamental and applied aspects of regeneration in higher plants. However, the use of these techniques is not confined to research alone. Tissue culture techniques are also being exploited by many commercial laboratories. Even many horticultural companies are setting up small units to multiply plants which are difficult to propagate by conventional means.

In this chapter, the general organization of a tissue culture laboratory and the basic techniques have been discussed.

1.1 TISSUE CULTURE LABORATORY

An ideal tissue culture laboratory should have at least two big rooms and a small room. One big room is for general laboratory work such as preparation of media, autoclaving, distillation of water etc. The other big room, the growth room, is required for keeping cultures under controlled light, temperature and humidity. The small room known as 'Inoculation Room' is meant for aseptic work and for storing autoclaved articles.

1.2 ORGANIZATION OF LABORATORY

Plant tissue culture is commonly used to describe the *in vitro* and aseptic growth of any plant part on a nutrient medium. This technology is based on three fundamental objectives; 1) the part or explant must be isolated from the rest of plant body; 2) the explant must be maintained in controlled physically (environmental) and chemically (nutrient medium) defined conditions; and 3) asepsis must be maintained. It is required for asepsis to maintain a high degree of cleanliness in the laboratory, whether the techniques are being used for simple propagation, as a method to study genetic, metabolic, or developmental changes in a model system, or for the creation of new plant variations *via.* genetic engineering. There are a number of basic facilities and a minimum level of organization that should be made available to the people working in the laboratory. The basic facility comprises of the following–

1. A washing area. or A general clearing Area.
2. A media preparation, sterilization area and storage area.
3. A transfer area (with laminar air flow bench) for aseptic manipulation. or inoculation room.
4. Culture of growth chambers. or Environmentally controlled incubators.
5. An observation and experimentation area.

This work plan may be extended to different sizes depending upon the type of research involved with plant tissue culture, *e.g.*, micropropagation, genetics, electron microscopy, biochemistry, molecular biology, production of secondary metabolities, etc. Organization of a small tissue culture laboratory for micropropagation and related research work is closed type system to avoid exposure to external environment which may create problem of contamination. Entry of personnel is restricted, as far as possible in the sterile areas. Materials and persons from exterior are not allowed to enter in the sterile areas directly. All these precautions are necessary to maintain asepsis.

1.2.1 Washing Facility

The area for washing facility should include the following–

1. Large sinks, drainage racks, demineralized, distilled and double distilled water (Distillation Unit).
2. Ovens for drying the glassware after washing, and storage cabinets for dried glassware.
3. Generally glasswares are cleaned by detergents, washed thoroughly under running tap water, rinsed with distilled water, oven sterilized and stored. Some glasswares may require cleaning with ultrasonic bath for removing particles and stains.

***Glasswares*:** Usually good quality, heat resistant, borosilicate glasswares are used and good stock of glassware is required to run a laboratory. A large number of glasswares are utilized at different levels and it takes time in re–circulating the glassware. All types of glasswares are required in a functional plant tissue culture laboratory according to experimental requirement, specially conical flasks, culture tubes, measuring cylinders, Petri dishes (also disposable pre–sterilized plastic Petri dishes), pipettes and culture vessels of different shapes and sizes are required in large quantities.

1.2.2 Medium Preparation Room

A media preparation room should have ample benches (working tables) and storage space for glassware, culture vessels, cabinets, balance, pH meter, refrigerator, deep freezer, autoclave, magnetic stirrer etc. Distilled water plant is essential and can be kept either in washing or medium preparation room. The most common method of water purification for tissue culture use is a de–ionization treatment followed by one or two glass distillation. The deionization treatment typically removes most of the ionic impurities, but particulate matter, especially organic compounds, pyrogens and non–ionized leachate from the ion–exchange resin, as well as microbes and gases, are not effectively removed by demineralization. Proper distillation will effectively remove large organic molecules, micro–organisms, pyrogens and other ionic and nonionic contaminants.

1.2.3 Transfer Area (Inoculation Room)

Though the initial research on tissue culture is carried out in an open laboratory bench, but a sterile transfer room with a laminar air–flow bench should be made available for aseptic manipulations.

The maintenance of asepsis is essential in all manipulations. A small room can be fitted with overhead ultra violet light and a positive pressure ventilation unit with a bacteria proof High Efficiency Particulate Air (HEPA) filter. This chamber or laminar flow bench should be provided with gas, electricity and compressed air out–lets for use during transfer. The area should be cleaned regularly, without creating dusty atmosphere.

1.2.4 Culture Facility (Growth Rooms)

Plant tissue cultures, cell suspension in liquid or static cultures on solid media, should be incubated under environmentally controlled conditions of temperature, humidity, illumination, photoperiod and air circulation. All of these factors can affect growth and differentiation response of tissues during culture. Anthers and protoplasts are especially sensitive to environmental conditions. Small growth rooms should meet these requirements. Plant growth chamber should be equipped with air conditioners, temperature controller, timer switches, photoperiodic controller, humidifier and illuminated racks.

1.2.4.1 *Light*

Light is fixed in the roof of equipment or in shelves. Light is generally provided by commercially available light sources like cool white fluorescent tubes fixed in the shelves of tissue culture racks. Usually a light intensity of 2000–4000 lux (or 30 µmol $m^{-2}s^{1}$) is used. The duration of light and dark cycle is adjusted as per requirement, usually 16 hours light cycle (photoperiod) is given. Now a days warm fluorescent tubes are also available which provide wide spectrum as compared to cool white fluorescent tubes and mixing of incandescent lights is not required with former tubes. Light intensity can also be regulated by photoperiod simulators. Thus, light quality, intensity and period is controlled and regulated by the instruments as per set parameters.

1.2.4.2 *Temperature*

In modern equipment, temperature is precisely regulated by good quality (platinum) temperature sensor. In all cases, air conditioning units provide the cooling. It is always advisable to keep one spare compressor unit, for emergency, to avoid delay in repairs and damage to cultures. Usually temperature of 22–28°C is used for growing plant tissue cultures. Growth room should be provided with air circulating fans to have uniform temperature in each and every corner of the room.

1.2.4.3 *Humidity*

Humidity inside the growth chamber is provided by humidifier (a mist generating system) and controlled by humidistat. Usually 60% RH (relative humidity) is used to maintain healthy growth. Low humidity may cause early drying of medium while high humidity may cause fungal growth in the environment and on various articles. Thus, in a growth chamber, light, temperature and humidity are precisely controlled and cultures are grown in a controlled environment. All the controls are regulated through a control panel (Fig.1.1).

1.2.5 Culture Conditions

The environment of the cultures includes, nutrient solution (medium) and physical culture conditions in which cultures are allowed to grow.

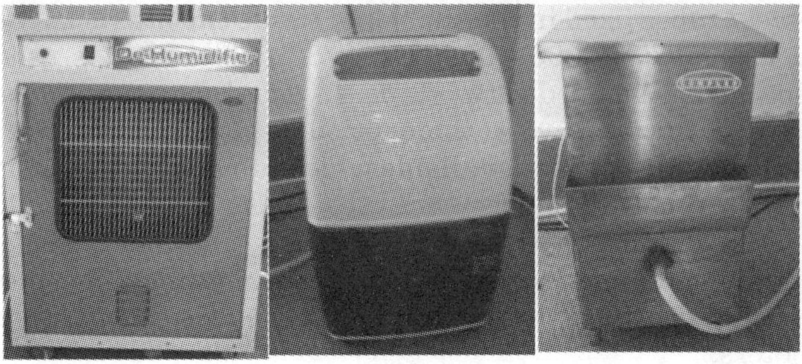

Fig. 1.1: Dehumidifier (left) and humidifiers (middle and right)

(a) ***The culture medium*:** Plant cells can be cultured on simple nutrient medium as compared to animal cells, which require complex substances. The most commonly used media are Murashige and Skoog (1962) and B_5 of Gamborg *et al.* (1968). Essentially all the media contain similar formulations in different composition. The names of these media are known after the scientists who formulated them.

(b) ***Physical Environment*:** Culture conditions have direct bearing on the growth of the tissues. Following conditions are generally used.

- *Temperature of incubation*: 22–28°C.
- Illumination 0–5000 lux, with a photoperiod of 8–16 hrs.
- Relative humidity, 60–80% RH.
- *Subculture of tissues*: 2–8 weeks for static cultures and 1–2 weeks for cell suspension cultures.

Composition of the culture medium and the culture conditions are strictly maintained for the cultures of a given species. Modifications in any of these (singly or in combination) may cause variation in growth and metabolism of the cultured tissues during different experiments.

In case of liquid cultures the vessels are agitated at about 80–120 rpm (revolutions per minute) on a gyratory shaker. There should be sufficient space in the culture room to place shakers of various types, bioreactors and shelves in sufficient numbers for static cultures.

Chapter 2

Tools and Techniques for Plant Tissue Culture

EQUIPMENTS USED IN PLANT TISSUE CULTURE

Following equipments are generally required in a plant tissue culture laboratory.

2.1 pH METER

The hydrogen ion concentration of most solutions is generally low. In 1909 Sorenson introduced the term pH as a convenient way of expressing hydrogen ion concentration. pH of a solution is strictly defined as the negative logarithm of the hydrogen ion activity but in practice usually hydrogen ion concentration is taken.

$$pH = -\log_{10} [H^+] = 7$$

pH meter is an essential–equipment for measurement and adjustment of pH of the nutrient medium (Fig. 2.1). The pH of pure water is 7 at 25°C. Generally glass distilled water is used for the preparation of culture medium. However sometimes buffered solutions may be used to keep the pH of the medium constant.

2.1.1 Measurement of pH

An approximate idea of the pH of a solution can be obtained using indicators. These are organic compounds of natural or synthetic

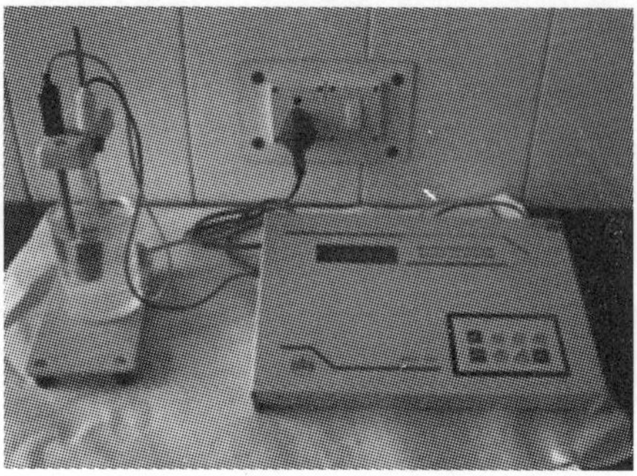

Fig. 2.1: A digital pH meter

origin, whose colour is dependent upon the pH of the solution. Indicators are usually weak acids, which dissociate in solution. A standard pH meter has two electrodes, one glass electrode for measuring pH and the other calomel reference electrode which is filled with saturated KCI solution. In modern sophisticated pH meters, both types of electrodes are combined in one.

2.1.2 Applications

1. To adjust pH of different solutions, culture media and preparation of buffers.
2. Determination of pH of cells (cells sap) and in analytical techniques.
3. To monitor pH of the medium in a bioreactor.

2.2 HOT AIR OVEN

It is necessary for drying the washed glasswares. For this purpose, a number of enameled trays of different sizes are required for keeping glass goods inside the oven (Fig. 2.2).

2.3 REFRIGERATOR

It is essential for storing various thermoliable chemicals like vitamins, hormones, amino acids, casein–hydrolysate, yeast extract, coconut milk etc. Stock solutions of salts are also kept in refrigerator to prevent contamination (Fig. 2.3).

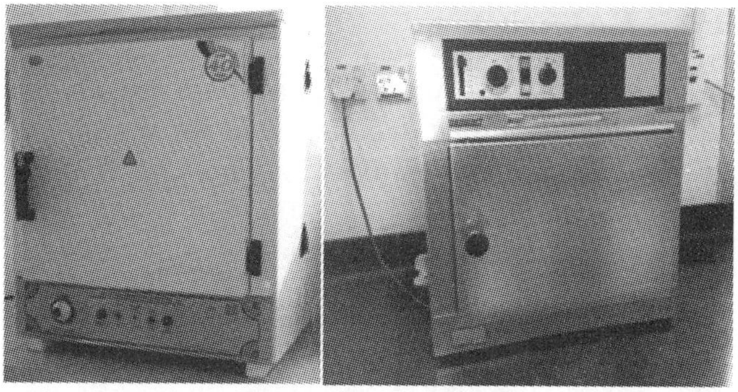

Fig. 2.2: Hot air ovens

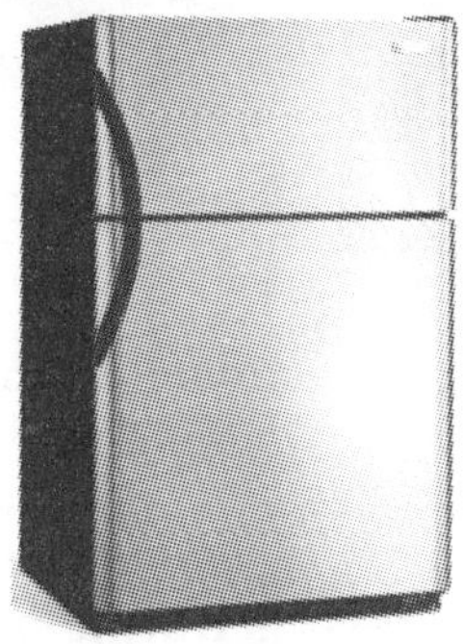

Fig. 2.3: Refrigerator

2.4 WATER DISTILLATION PLANT

Single distillation and a double distillation water plants are essentially required. Big plastic containers are also required for storing the distilled water (Fig. 2.4).

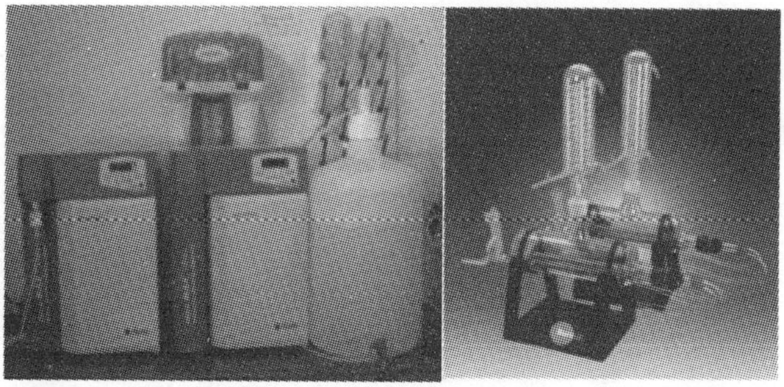

Fig. 2.4: Types of water distillation plant. RO Water system (*Left*) and Double distillation unit (*Right*)

2.5 WEIGHING BALANCE

Three types of weighing balances *viz*., pan balance, chemical balance and electronic digital balance are required for weighing chemicals, sugars, agar–agar and other adjuvants (Figs. 2.5 & 2.6).

Figs. 2.5 & 2.6: Electronic weighing balances

2.6 VACUUM PUMP

It may be required for filtering liquid media, sugar solution etc through filter apparatus using air suction.

2.7 WORKING TABLES

These are necessary for preparation of medium and other laboratory work.

Fig. 2.7: Different types of working table

2.8 HEATING MANTLE

It is needed for heating or warming the medium to dissolve agar or to melt the agar-gelled medium (Fig. 2.8).

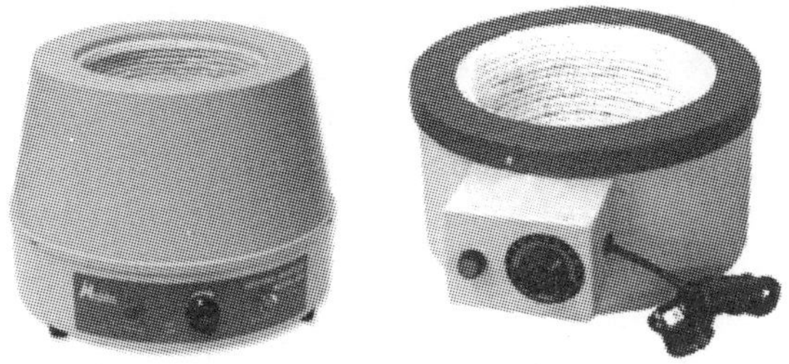

Fig. 2.8: Different types of heating mantle

2.9 MICROTOME

A microtome is needed for sectioning the cultured tissues/explants for their histological studies (Fig. 2.9).

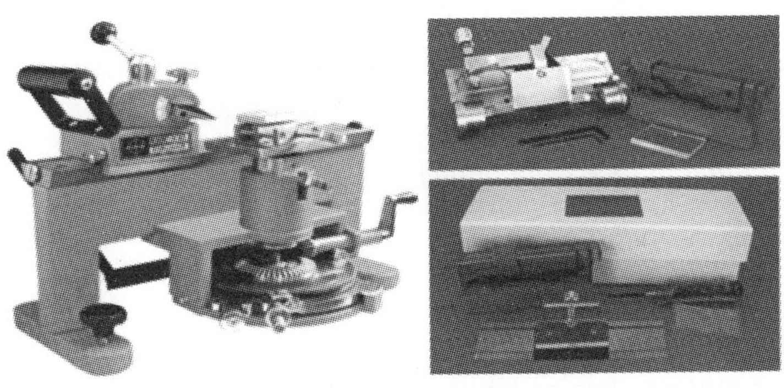

Fig. 2.9: Different types of microtome

2.10 WOODEN RACKS

Wooden or metallic racks (Fig. 2.10) of suitable sizes are required for storage of various chemicals.

2.11 LAMINAR AIR FLOW CABINET

Laminar air flow cabinet (Fig. 2.11) is the most suitable, convenient and reliable instrument for aseptic work. It allows one to work for a

Fig. 2.10: Different types of wooden racks

longer period which is not possible inside the inoculation chamber. Long hours of work inside the inoculation chamber may also cause suffocation and needs the interruption of work.

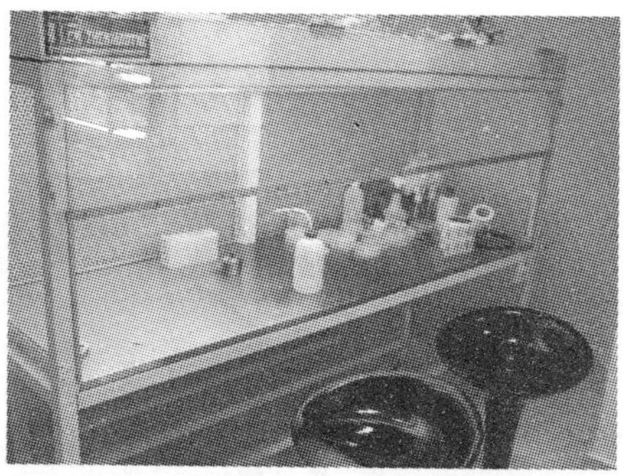

Fig. 2.11: A Laminar air flow

One can work easily for a longer period on the table of laminar air flow. Laminar air flow has a number of small blower motors to blow air which passes through a number of HEPA (high efficiency particulate air) filters. Such filters remove particles larger than 0.3 µm. The ultraclean air which is free from fungal and bacterial contaminants, flows at the velocity of about 27 m^3/minute through the working area. All contaminants are blown away by the ultraclean air and thereby an aseptic environment is maintained over the working area. Before starting work, laminar air flow is put on for 10-15 minutes. The flow of air does not put out the flame of a spirit lamp. Therefore, a spirit lamp can be used conveniently during the work.

2.12 GLASS GOODS

Different types of glass goods (Fig. 2.12) are used to culture plant tissues. The conventional and some specific glass goods are required for culture work. Glass goods should be of coring or Pyrex or similar boro–silicate glass. Measuring cylinders, conical flasks, pipettes, beakers are required for preparation of media. Plant tissues are grown in wide–necked Erlenmeyer conical flask (100 ml, 150 m1, 250 ml etc.), culture tubes (25 mm in diameter and 150 mm in length), Pretri–dishes (50, 90, 140 mm in diameter), screw–capped universal bottles (20 cm^3 capacity). Sometimes jam bottle, milk bottle may also be used. Particular care must be taken to ensure that glass goods are properly cleaned before use.

In tissue culture two ability to accurately and reproducibly measure and transfer small volumes of liquids in critical for obtaining useful results. For volumes less than 1 ml, the most common method for measuring liquid volumes involves the use of a device known as a pipetman or micro-pipette.

The pipetman can be both fixed and variable volumes commonly used pipetman are P 1000, P 200, P 20, P 10, etc. P 100 are useful for

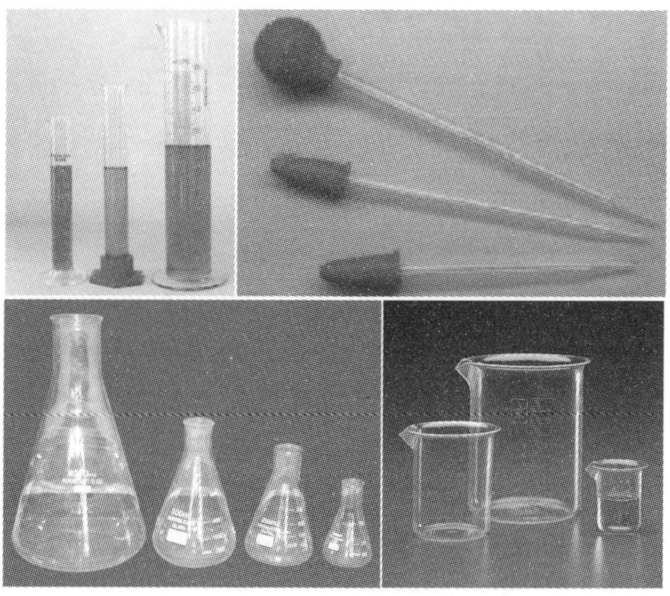

Fig. 2.12: (A) Different types of glass goods

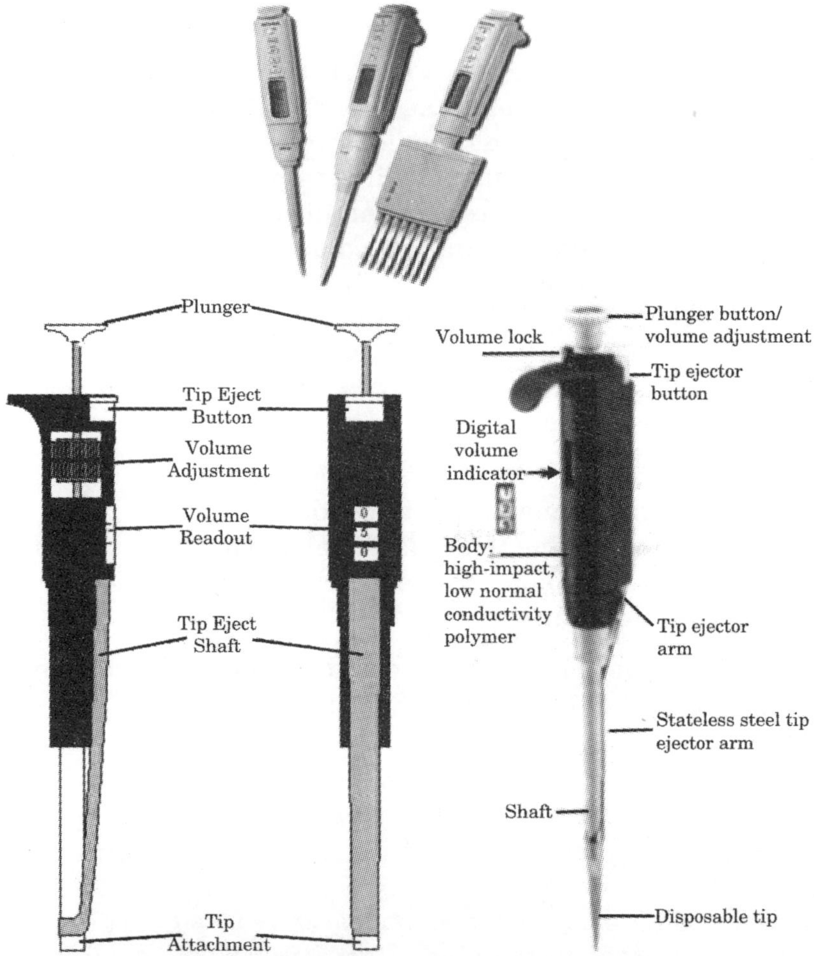

Fig. 2.12: (B) Parts of a Micropipette

volumes from 200 to 1000 ml. P 200 are useful for volumes from 20 to 200 ml. P 20 are useful for volumes from 0.5 to 20 ml. For beginners fixed volume pipetman are more useful. It is important to ascertain whether the correct pipetman is chosen for measuring the volume one needs.

The traditional method of cleaning new or dirty glass goods is to soak these in soap water followed by brushing and washing well with tap water and finally rinsing with single distilled water. These are dried in the hot air oven and then the clean glass rods are stored in a dust–proof cupboard or drawer. In order to autoclave the culture medium and to culture the plant material, culture vessels particularly, culture flasks and culture tubes must be fitted with cotton plugs which exclude microbial contaminants, yet allow free gas exchange. For this, tightly rolled plugs of non–absorbent cotton wrapped in gauge cloth may be used.

When in position the exposed part of each plug and the rim of the culture vessel should be covered by brown paper or a cap of aluminium foil. This will keep the plug and vessel rim free from dust and will protect the plug from wetting during autoclaving. In some laboratories, pre–sterilized, disposable plastic wares are used in order to culture plant tissues. Some of these plastic wares are autoclavable. For the sterilization of set of compounds or enzymes for protoplast isolation a specially designed glass made bacterial filter (Fig. 2.13) or an autoclavable plastic made bacterial filter is used.

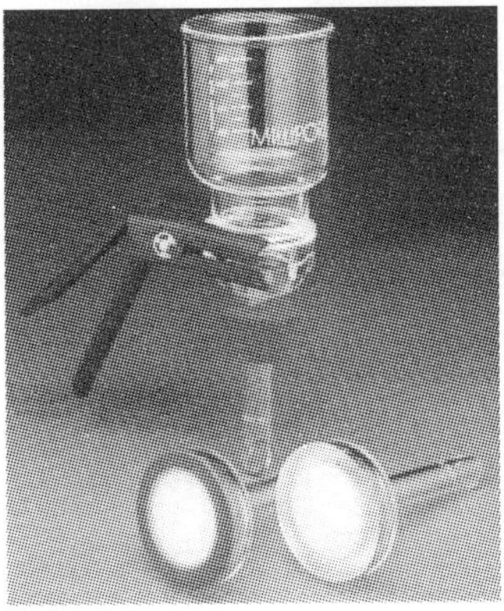

Fig. 2.13: Specially designed glass made bacterial filter

A small spirit lamp made of glass may be required for the flame sterilization of surgical instruments using methylated spirits.

2.13 SURGICAL INSTRUMENTS

Surgical instruments routinely used in tissue culture work include various sizes of scalpel, forceps, spatula, scissors, etc (Fig. 2.14) described as under.

Table 2.1: Measurement of Surgical instruments

Instrument	*Measurement in length (cm)*	*Size no.*	*Use*
1. Scalpel	20.32, 25.4 etc.	–	For cutting the plant material
2. Surgical scalpel–			Do
(a) Holder	–	3,	
(b) Disposable blade (arrow headed)	–	411	
3. Forceps (without corrugated marking at the pointed end)	20.32, 25.5	–	For holding the material
4. Jewellery fine forceps	–		For peeling leaf epidermis

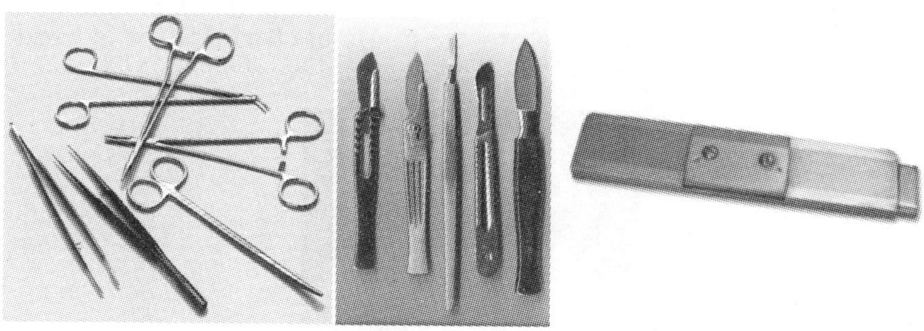

Fig. 2.14: Surgical instruments routinely used in tissue culture work

2.14 AUTOCLAVE

Sterilization of material is carried out in an autoclave increasing moist heat (121°C) due to increased pressure inside the vessel (15–22 per pounds per square inch or 1.02 to 1.5 kg/cm^2) for 15 minutes for routine sterilization. Moist heat kills the microorganisms and makes the material free from microbes. Vertical and horizontal models of autoclave (Figs. 2.15 and 2.16) are used to sterilize medium, glassware and surgical tools used in plant tissue culture. The same equipment is used in hospitals to sterilize gauge, cotton, tools and lines etc. The time of sterilization depends on the volume of media (Table 3.1).

Fig. 2.15: Vertical autoclave

Fig. 2.16: Horizontal autoclave

2.14.1 Construction

Autoclaves of different sizes from 5 liters to several hundred liters capacity are available in horizontal or vertical designs. As shown in Figs. 2.15 and 2.16, an autoclave has a body, an internal (or external for small sized autoclaves comparable to house–hold pressure cooker) heating system and a container to hold material. Its cover is fixed with pressure gauge, safety valve, pressure release valve etc. Lid is tightened with the help of screws and a gasket seals the body and lid. A jacket, paddle lifter timer and indicator, etc. are also provided with large sized autoclaves. Autoclaves may be constructed of aluminium, mild steel, stainless steel or gun metal. Industrial autoclave can accommodate large trolley containing huge number of glassware.

Figure 2.17: (a) shows an autoclave with its longitudinal axis arranged horizontally, provided with a jacket through which a heat exchange fluid can be circulated. The fluid is introduced through an inlet from a source (not shown) and leaves through an outlet. The autoclave has an inlet through which the monomer and other necessary reagents are introduced and an outlet through which the unreacted monomer is removed at the end of the polymerization. The autoclave also has an outlet controlled by a valve through which the polymer is discharged. The monomer inlet and polymer outlet are situated at opposite ends of the autoclave.

The apparatus 10 shown in the Fig. 2.17 (b) includes a steel drum which is mounted for rotation, about a shaft within a pressure

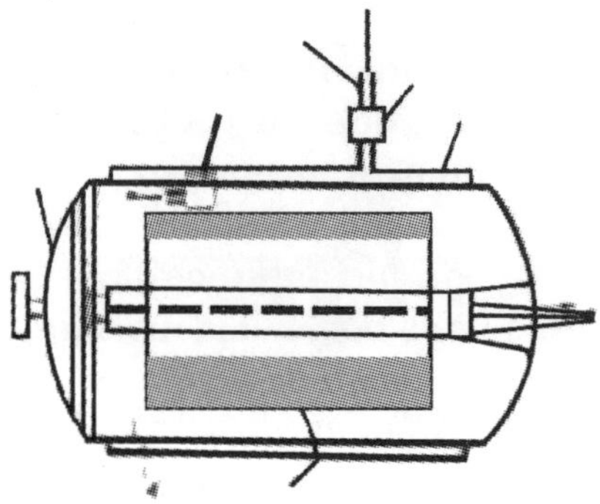

Fig. 2.17a Auto clave

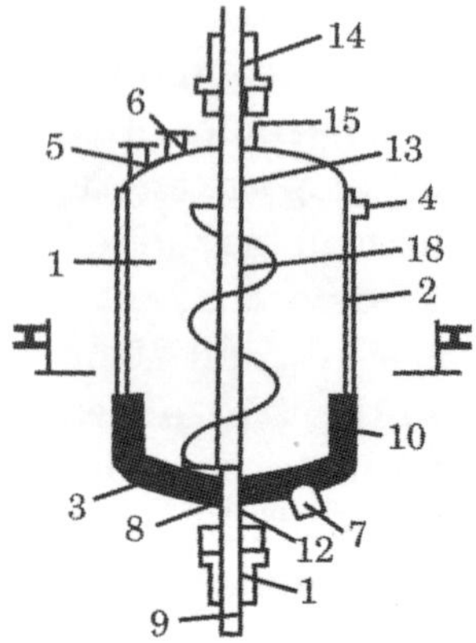

Fig. 2.17b Steel drum

vessel. Preferably pressure vessel is cylindrical (as illustrated), although it may be of any other suitable shape. An endless belt is arranged around a series of eight rollers such that it travels into vessel through an entry aperture around about 80% of the circumferential surface of drum and then out an exit aperture 18. One of the rollers 16 is adjustable in a known way for adjusting the tension in belt, also shaft or one of the rollers is driven. A fabric is fed to the apparatus such that it is sandwiched between endless belt and drum. A textile material (or wrapper) for example of cotton which is made wet, can also be fed to apparatus such that it is sandwiched between fabric and the surface of drum. The entry and exit apertures and for vessel each include a sealing arrangement (which may include rollers) to minimise the leakage of compressed air from vessel. Suitable seals are known in the art and are thus not further described herein. Drum 12 includes a suitable arrangement (shown substantially by block 22) for heating its surface up to, for example, a temperature of about 180°C. It will be evident from Fig. 2.17(a) that a clockwise rotation of drum will convey fabric 19 and the wet cotton wrapper through a treatment zone between the heated outer facing surface of drum and the inner facing surface of belt, which zone extends around the drum starting from about where the belt contacts the surface of drum after entry into vessel and finishing about where it leaves the drum just prior to its exit from vessel.

Belt 15 may be made of an impermeable silicone material. As a fabric 19 is transported through the vessel, it is compressed between belt and drum sufficiently to ensure the fabric retains its flatness, that is, to ensure that a desirable fabric finish is maintained. This compression is provided by tension in belt assisted by the force which the pressurized atmosphere within vessel exerts on the belt. That is, it is to be understood that belt is tensioned, but this tensioning and the consequential lateral pressure that the belt exerts on the fabric is relatively low in that the belt's tension is required only to exert reasonable lateral pressure to maintain the fabric flatness. Simultaneously, the interaction between the heated surface of drum and wet surface of wrapper causes a rapid evaporative/condensative heat transfer process to occur which establishes a saturated (or superheated, depending oil the temperature and the amount of moisture) steam atmosphere within the treatment zone.

Use of a wet cotton wrapper, which gives a saturated regain to the fabric, is not essential and as an alternative wrapper can be omitted and the surface of the fabric 19 which contacts the heated surface of drum can be moistened prior to entering the apparatus. For example, free water can be applied to the surface of fabric by a spray (not shown). Alternatively, steam may be introduced into the treatment zone via perforations in the drum, or from a steam chamber arrangement provided within the pressure vessel 14 *via.,* a permeable belt.

2.14.2 Operation

Place the materials (wrapped in aluminium foil or paper or in metal box) and glasswares containing medium (plugged with non–absorbent cotton and covered with aluminium foil) in the metallic bucket. Check water level for appropriate level, tighten all the screws, and switch–on the current. Allow the steam to pass freely from release valve for 5 minutes and then close the valve. After attaining a pressure of 15 psi, count 15 minutes for sterilization and then switch–off the current. Pressure is maintained by safety valve. Modern autoclaves are fitted with temperature and time control devices that can automatically control the period for sterilization and then switch–off automatically.

For large sized vessels and large volume flasks containing high amount of liquid, duration for sterilization should be increased accordingly. Fig. 2.18 shows the relationship between volume of the solution and duration for its sterilization at 15 psi. The vessels should not be filled more than 1/3 of its capacity for proper sterilization.

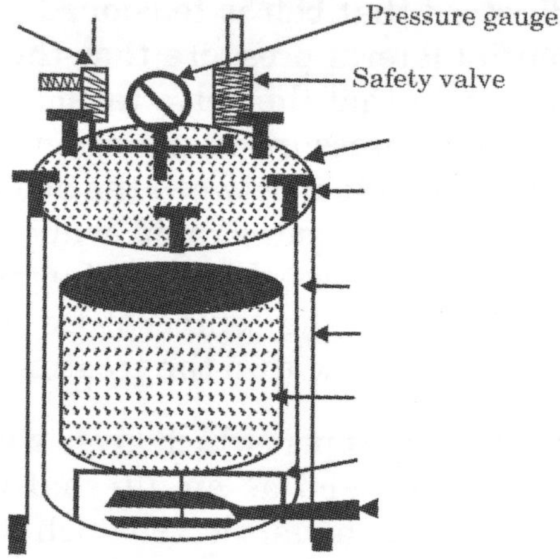

Fig. 2.18: A metallic bucket

2.14.3 Precautions

- Check water level each time, the heating elements should remain immersed in the water.
- Check spring of safety valve frequently and clean opening whenever necessary.
- All the electrical equipment should be properly earthen to avoid electric shock.
- Opposite screws of lid should be tightened simultaneously.
- Do not over tighten the screw to avoid damage to the gasket.
- Use permanent marker to mark your medium.

2.15 MICROSCOPES

Simple, compound, inverted or zoom stereoscopic binocular/trinocular microscopes are essential for various purposes. Some of the microscopes should be fitted with a camera for photomicrography.

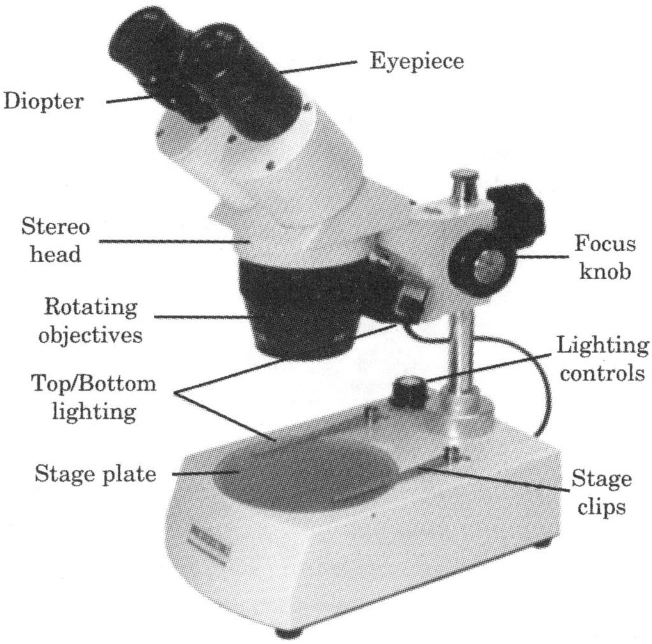

Fig. 2.19: A zoom stereoscopic binocular microscope

2.15.1 Electron Microscopy

Electron microscopy permits a detailed study of sub–cellular organelles as its resolving power is much greater than that of the light microscope. Max Knoll and Ernst Ruska in 1931, at Technical University in Berlin, constructed electron microscope (EM). In the EM, streams of electrons are detected by an electrostatic or electromagnetic field in the same way that a beam of light is refracted by a lens. Electron beam is generated by heating a filament in vacuum, which are accelerated by a potential and shows properties similar to light (A = 0.005 nm of electrons and 550 nm for light). Though it appears to be similar there are great differences in light and electron microscopes, the principal being the image formation (Fig. 2.20). In electron microscope, image is produced by electron scattering. Electron dispersion is a function of the thickness and molecular packing of the object and depends especially on the atomic number of the atoms in the object. The higher the atomic number, the greatest is the dispersion.

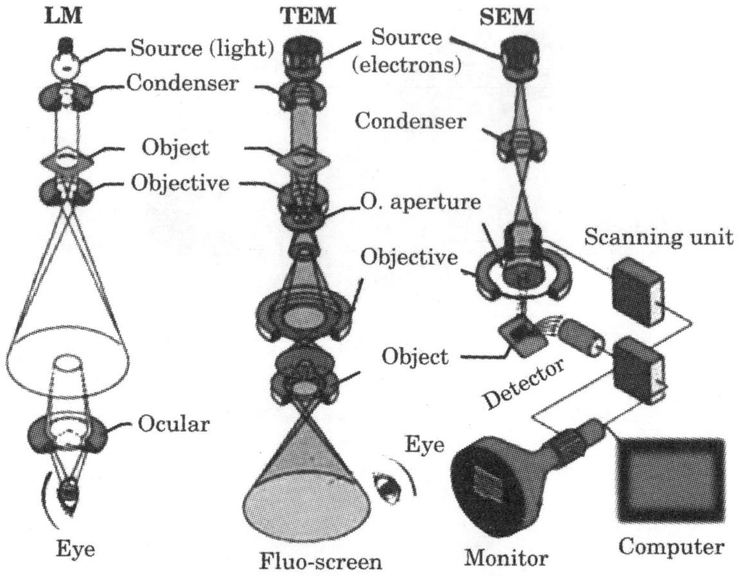

Fig. 2.20: Diagrammatic representation of image formation in electron microscope and light microscope

There are three general types of electromagnetic lenses. The one is placed between the source of illumination and the specimen. This focuses the beam of electrons on specimen and functions in a similar manner as the condenser in light microscope. The other two lens systems are on the opposite side of specimen which magnifies the image in similar way as objective and ocular in light microscope. The final image is produced on a screen coated with a phosphorus compound which fluoresces upon irradiation by electrons. These images are recorded on photographic films. Air molecules in the microscope interfere with the movement of electrons. To prevent this, a high vacuum (10^{-4}–10^{-6} mm Hg) is created inside the microscope. In light microscope, magnification is largely determined by the objective lens and the maximum magnification of 100–120 can be obtained. This is multiplied by ocular lens by a factor of 5–15, reaching a total magnification of 500–1500x.

The resolving power of Transmission Electron Microscope (TEM) is very high. The image generated by objective can be multiplied several hundred times by projector coil, *e.g.*, 100 objective 200x, projector coil =20,000x. In TEM, this can be reached up to 10,000x. Electron microscope has a greater depth of field as compared to light microscope. This property of EM is used to develop three dimensional images of cell organelles as ribosome and protein structure.

The preparation of specimen is similar to dehydration used for light microscopy. The sections prepared for EM should be thin (less than 0.5µ or 500 nm) otherwise material will appear completely dark. Such thin sections are prepared with the help of ultra microtome, fixed with glass or diamond knife. To cut sections, material is fixed in osmium tetra oxide, processed through a dehydration series and embedded in plastid or resin (araldite). Sections are placed on small grids specifically used for EM. It is difficult to handle ultra thin sections so they are placed on copper grids (400 aperture/inch square). The grids are placed in labeled boxes and can be stored till observed in EM and can be used again and again without damaging the material. Dehydrated materials are placed in vacuum chamber and vapor of heavy metals like platinum, chromium or palladium is directed from a filament of incandescent tungsten. The vaporized material is deposited on one side of the surface of elevated particles. This creates a shadow on the other side (Shadow casting). Height of the material is determined by length of shadow. In photographs, such specimens appear in three dimensions, which is not possible with other technique. Freeze fracturing (fracture of frozen material) is used to see fine details of replicated material. There are different techniques to prepare materials for sectioning and for staining to increase contrast. Special techniques are used for bacteria, viruses, proteins and nucleic acids. EM is used to study the fine details of a cell, compare the clones, development of plastids and cell wall formation.

2.15.2 SEM

Scanning Electron Microscope (SEM) provides surface views of whole structure of specimen (Fig. 2.21). Normally, specimens are coated with a thin film of metal under vacuum. As compared to shadow casting, complete coating is essential because title scanning beam produces a charge on the uncoated biological materials which causes disturbance of image.

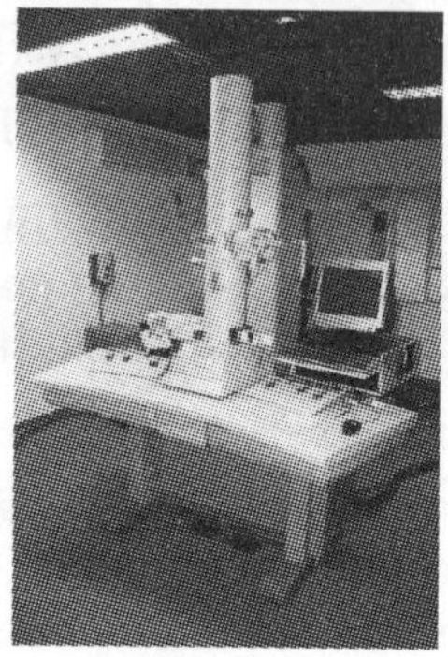

Fig. 2.21: Scanning electron microscope (SEM) (50–100Å.)

The metal coated specimen is scanned with a beam of electrons which strikes on the specimen and emit secondary electrons. These electrons pass through a cathode tube and image is produced on the screen of Cathode Ray Tube. In plant system, SEM is used to see the nature of appendages, pollen genesis in plant tissue culture. It is also used to see structures of metal crystals and in forensic sciences.

2.15.3 Light Microscopy

Bright field microscopy is absolutely indispensable tool for cell biologists. Light microscope (Fig. 2.22) is required for routine observations of cells, cellular differentiates and pigmentation. According to wave theory, light is propagated from one place to another as wave travelling in a hypothetical medium. Light waves are described in terms of their amplitude, frequency, and wave length. Amplitude is the maximum, displacement of light path from the position of equilibrium. Frequency of light is the number of complete cycles occurring in a second. The property of light wave is inversely associated with the frequency is the wavelength and is defined as the distance between corresponding points on a wave or distance between two successive peaks or crests. The velocity of light is about 186300 miles or 3×10^{15} km per second.

A good microscope has not only good magnifying power but also good resolving power to provide finer details of the object. Thus, the basic difficulty in designing a microscope is not the magnification, but the ability of lens system to distinguish two adjacent points as distinct and separate. This ability is known as resolving power of the microscope. The resolving power of a microscope depends upon the wavelength of light and numerical aperture (KNEE). The minimum resolvable distance between two luminous points (V) is given by the following formula–

V= 0.16/N.A. where, = Wavelength

Thus, shorter the wavelength of light used and lower the KNEE, greater is the resolving power. The limit of resolution of a microscope is approximately equal to 0.5/N.A, which for a light microscope is approximately 200 nanometers (nm).

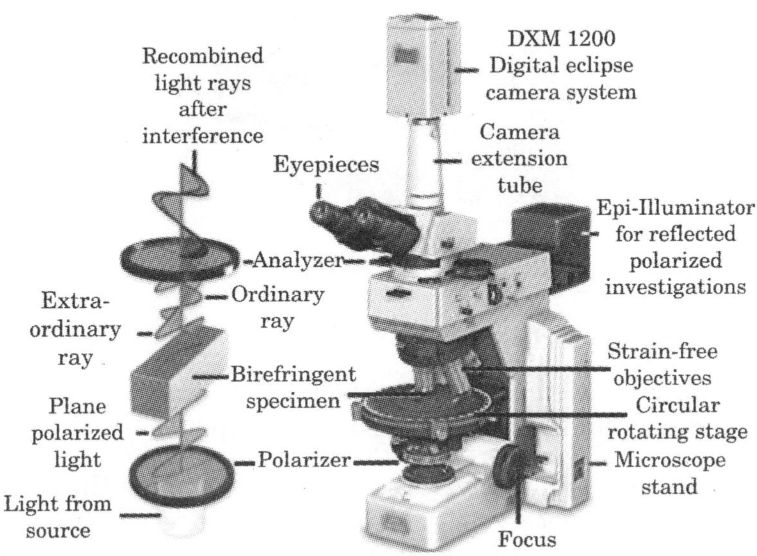

Fig. 2.22: Microscope with photomicrography attachment

2.16 COLORIMETER

The most commonly used method for determination of biochemical compounds is colorimetry using a colorimeter or spectrophotometer

(Fig. 2.23). It uses the property of light such that when white light passes through a coloured solution, some wavelengths are absorbed more than others. Hyaline solution can be made coloured by specific reactions with suitable reagents. These reactions are generally very sensitive to determine quantities of material in the region of millimole per liter concentration. The big advantage is that complete isolation of the compound is not necessary and the constituents of a complex mixture such as blood can be determined after little treatment. The depth of colour is directly proportional to the concentration of the compound being measured, while the amount of light absorbed is proportional to the intensity of the colour and therefore, to the concentration.

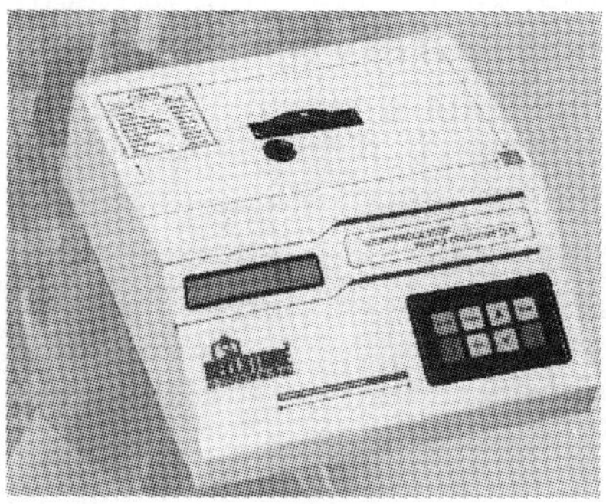

Fig. 2.23: Colorimeter

2.16.1 Source of Radiation

A lamp is usually used as the source of radiation. The wavelength of light depends upon the quality of source lamp.

Colorimeter is an analytical device to determine the amount of an unknown substance in a solution. The device works on optical properties of the substance. Basically the amount of light of a particular wavelength absorbed by that solution, and relate solute concentration to the absorbance is measured. A simplified schematic presentation of the instrument is as follows:–

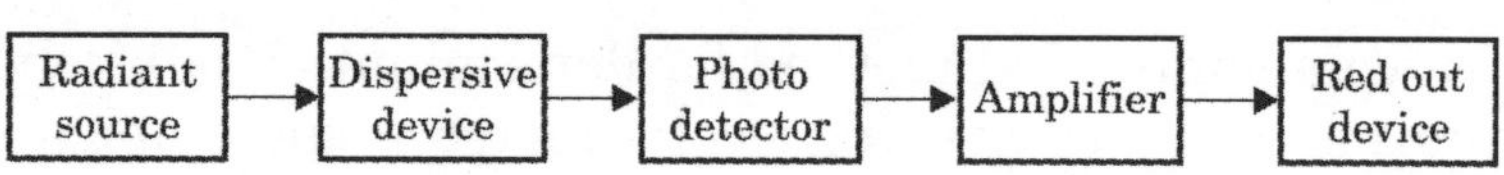

2.16.2 Applications

***Measurement of concentrations*:** Colorimeters or spectrophotometers are widely used for quantitative measurements of substance concentration in the solutions by measuring absorbance at the optimal wave length. This is used to determine concentration in fractions obtained in column chromatography, TLC and in other solutions.

***Absorption spectra*:** Identification and structure evaluation of various isolated pure compounds can be done by UV–Visible spectrophotometry. All the compounds have very specific absorption spectra depending on the molecular structure. Comparison of spectra helps in identification of compounds. The wavelength at which maximum absorption takes place is represented as αmax of that compound in a given solvent.

***Enzyme kinetics*:** In enzymatic reactions, changes in absorbance are recorded to determine the reaction velocity and concentration of the product formed/substrate utilized. Spectrophotometers coupled with microprocessor or computer can store data, figures and are helpful in comparing the effect of different variables.

2.17 CENTRIFUGE

A centrifuge (Fig. 2.24) is an instrument which produces centrifugal force by rotating the samples around a central axis with the help of an electric motor. Centrifuges can be categorized as the clinical type

Fig. 2.24: Centrifuge

(5–10,000 rpm) refrigerated high–speed centrifuges (910,000–20,000 rpm) and ultra–centrifuges (20,000 to 80,000 rpm). With increase in rpm, the friction of rotor with air produces so much of heat that they have to be run under refrigeration (so called refrigerated centrifuge) and both refrigeration and vacuum are used in ultracentrifuge, which runs at very high rpm, For these high speeds, even they has to be made of special metal to withstand the great force.

There are two types of rotors, angle head and swing. In the former, the samples are kept at an angle of about 30° to the vertical axis whereas in the later the samples while spinning are horizontal. Simple calculations show that for the same radius, the swinging bucket method produces more gravitational force. Ultracentrifuges are of two types–analytical and preparative.

2.17.1 Analytical Model

This consists of rotors and tubes, called cells. The instrument is designed to allow the operator to follow the progress of the substances in the cells, while the process of centrifugation is in progress. By estimating sedimentation velocity during the process, the molecular weight, purity etc. can be determined.

2.17.2 Preparative Model

This is used for purification of the components of macromolecules or other substances and all determinations are made at the end of centrifugation. The instrument has no monitoring device, while large centrifugal forces are set for a fixed time period.

Centrifugation is the most widely used technique for separation of various metabolites and also used to separate non–miscible liquids during extraction of secondary metabolites, *e.g.*, water (aqueous), chloroform (organic) mixture. A wide variety of centrifuges are available, ranging in capacity and speed. During the process of centrifugation, solid particles experience a centrifugal force, which pulls them outwards *i.e.*, away from the center. The velocity with which a given solid particle moves through a liquid medium is related to angular velocity. The sedimentation coefficient is a characteristic constant for a molecule or a particle and is a function of the size, shape and density. It is equivalent to the average velocity per unit of acceleration. The unit Svedberg (s) is of the used with reference to centrifugation and is equivalent to a sedimentation coefficient of 10^{-13}s.

2.17.3 Applications

Centrifugation is applicable in analytical techniques, preparation of extracts, separation of non–mixable liquid of enzymes, inhibitors and removing particles (silica, etc.) from solvents. Centrifugation with heating and vacuum (suction is used for concentrating extraction instrument called simple concentrator) and rapidly removing solvents. All the enzymatic work requires refrigerated centrifuge to keep the samples cool during centrifugation and to protect enzymes from inactivation by heat.

2.18 CHROMATOGRAPHY

Chromatography (meaning 'colored writing') is a technique to separate molecules on the basis of differences in size, shape, mass, charge and adsorption properties. The term chromatography was used by Tswett, a Russian botanist, to describe the separation of plant pigments on a column of alumina. There are different types of chromatography but they all involve interactions between these components: the mixture to be 'separated, a solid pales, and 'a solvent. The magnitude of these interactions depends upon the particular method used. Usually Column chromatography is used to separate large quantities of compounds, whereas, paper chromatography (PC) or thin–layer chromatography (TLC) in one or two dimensions, is used for analytical work.

The mobile phase can be a gas or a liquid, whereas the stationary phase can only be a liquid or solid. In liquid column chromatography (LCC), separation involves a simple partitioning between two immiscible liquid phases, one stationary and the other mobile, the process is called liquid–liquid (or partition) chromatography (LLC). When physical surface forces are mainly involved in the retentive ability of the stationary phase, the process is denoted liquid–solid (r adsorption) chromatography (LSC). In ion exchange chromatography (IEC), ionic components of the sample are separated by selective exchange with counter ions of the stationary phase. Use of exclusion pickings as the stationary phase brings about a classification of molecules based largely on molecular geometry and size.

2.18.1 Paper Chromatography

Principle: Cellulose in the form of paper sheets makes an ideal support medium where water is adsorbed between the cellulose fibres and forms a stationary hydrophilic phase.

The suitably concentrated mixture is spotted onto the paper, dried with a hair–drier and the chromatogram is developed by allowing the solvent to flow along the sheet. The solvent front is marked and after drying the paper, the positions of the compounds present in the mixture are visualized by a suitable staining reaction (Fig. 2.25). The ratio of the distance moved by a compound to that moved, by the solvent is known as the R_f value and is more or less constant for a particular compound, solvent system and paper under carefully controlled conditions of solute concentration, temperature and pH.

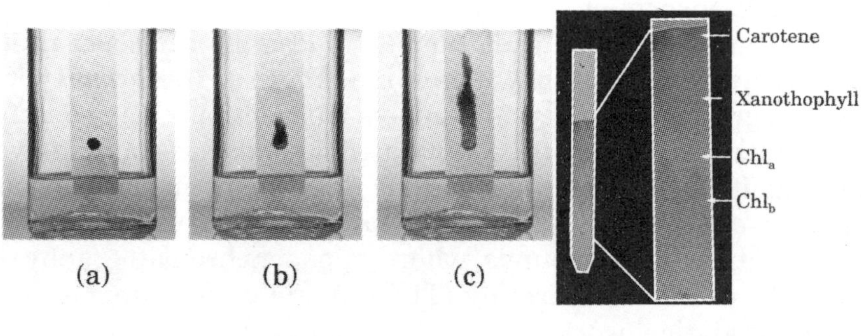

Fig. 2.25: Paper chromatography

Sample: Generally, alcoholic extracts of plant material with or without partial purification are used for chromatography. Biological materials should be desalted before chromatography by electrolysis or electrodialysis. Excess salt results in a poor chromatogram with spreading of spots and changes in their R_f values. It can also affect the chemical reactions used to detect the compounds being separated.

The sample (10–20 μl) is then applied to the paper with a micropipette or capillary.

Paper: Whatman No. 2 is the paper most frequently used for analytical purposes. Whatman No. 3 is a thick paper and is best employed for separating large quantities of material; the resolution is, however, inferior to Whatman No. 1. For rapid separation, Whatman Nos. 4 and 5 are convenient, although the spots are less well defined. In all cases, the flow rate is faster in the 'machine direction', which is normally noted on the box containing the paper. The paper may be impregnated with a buffer solution before use of chemically modified by acetylation. Ion exchange papers are also available commercially. For the separation of lipids and similar hydrophobic molecules, silica–impregnated papers are available commercially.

Solvent: This choice, like that of the paper, is largely empirical and will depend on the mixture investigated. If the compounds move close to the solvent front in solvent 'A' then they are too soluble, while if they are crowded around the origin in solvent 'B' then they are not sufficiently soluble. A suitable solvent for separation would, therefore be an appropriate mixture of 'A' and 'B', so that the R_f values of the components of the mixture are spread across the length of the paper. R_f value is defined as [R_f =the distance moved by solute/ the distance moved by solvent front]. This value is a constant for a particular compound under standard conditions and closely reflects the distribution coefficient for the compound. It is recommended that the developing chamber should be saturated with solvent by using filter paper lining inside the chamber.

2.18.2 Thin Layer Chromatography

Principle: Separation of compounds on a thin layer of adsorbing material is similar in many ways to paper chromatography, but has the added advantage that a variety of supporting media can be used so that separation can be done by adsorption ion exchange, partition chromatography, or gel filtration depending on the nature of the medium employed. The method is very rapid compared to paper chromatography and many separations can be completed within an hour. Compounds can be detected at a lower concentration that on paper as the spots are very compact. Furthermore, separated compounds can be detected by corrosive sprays and elevated temperatures with some thin layer materials, which of course is not possible with paper.

***Production of thin layer*:** The R_f value is affected by the thickness of the layer below 200 mm, and thickness of 250 µm is suitable for most separations. There are several good spreaders available in market, which can produce an even layer of required thickness by adjusting the thickness control new. Calcium sulphate is sometimes incorporated into the adsorbent to bind the layer to the plate a because of this. It is advisable to work rapidly once the adsorbent on various supporting material, such as glass, plastic and aluminum that are available commercially and these may be more convenient to us than trying to prepare plates in the laboratory.

***Development*:** It is essential to make sure that the atmosphere of the separation chamber is fully saturated with the solvent mixture, otherwise R_f values will vary widely from tank to tank. However, horizontal chambers of high performance TLC (HPTLC) are very useful and convenient as they require less time, solvent and no stabilization time. Development of the plate is usually by the ascending technique is very rapid (Fig. 2.26).

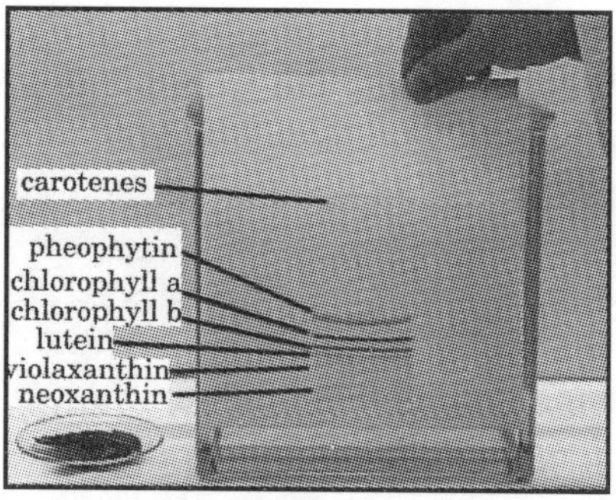

Fig. 2.26: Thin layer chromatography

2.19 THERMOMETERS

Temperature in growth room is measured by a room thermometer or a continuous rotary drum chart type thermometer. The U–shaped maximum–minimum thermometer is commonly used for determining diurnal maximum and minimum range of temperature in the culture room. During the dark period, temperature remains slightly (1–2°C) lower than light period (all bulbs and tube lights of the culture room increase the temperature). Therefore, chokes of the tube lights are fitted outside the culture room. The indicator needles of the maximum–minimum thermometer are moved by mercury column and they remain at that position until moved with the help of magnet. Their positions indicate the minimum and maximum temperatures in the previous 24h. After recording temperature indicators are reset to mercury level (Figs. 2.27 & 2.28).

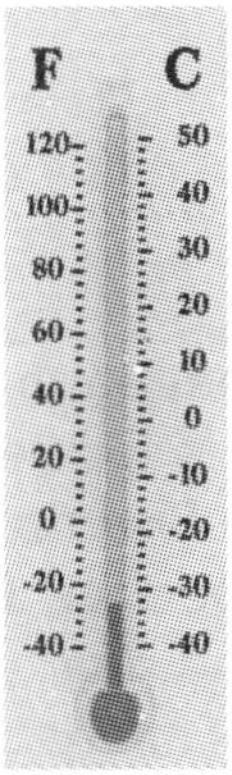

Fig. 2.27: Room thermometer

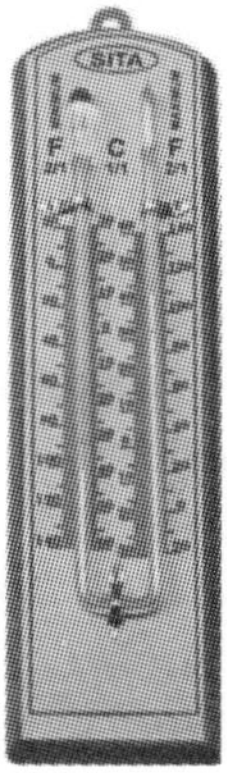

Fig. 2.28: Maximum–minimum thermometer

2.20 HYGROMETER

Dial type hygrometer (Fig. 2.29) is a convenient tool to measure humidity in the culture room. Humidifier is used to maintain 60% relative humidity (RH) in the culture room. Humidistate is a bimetallic thermocouple device to control and regulate the function of humidifier to maintain the humidity. Distilled water should be filled in the humidifier. The RH present in the culture room is measured by hair hygrometer. As the name suggests a chemically treated hair elongates with increased humidity and shortens with dryness (similar to mercury in thermometer). It is calibrated from 0 to 100% RH. At lower humidity, medium dries rapidly whereas at higher humidity chances of fungal growth over all surfaces and cotton plugs is increased. Therefore, about 60% RH is maintained in the culture room.

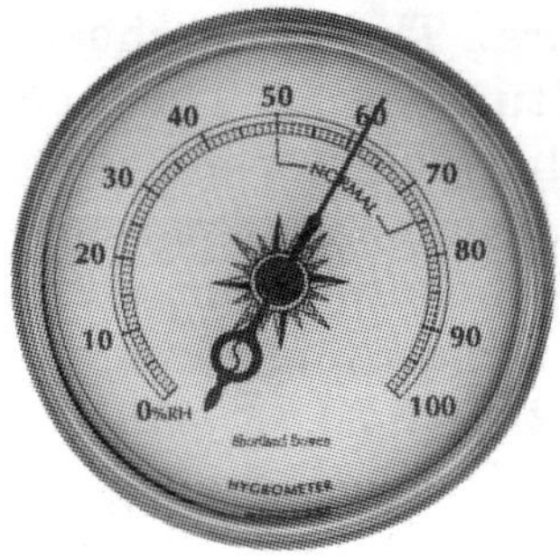

Fig. 2.29: Dial type hygrometer

2.21 LUX METER

Light is the form of radiant energy, *i.e.*, electromagnetic radiation of specific wavelength. Visible light as we perceive, is located in narrow wavelength region of spectrum between 380 to 760 nm. Light has dual characters, displaying both wave properties (refraction, diffraction, interference and polarization phenomena) and particle properties (light is radiated in discrete amounts of energy or photons).

Properties of light should be defined either as irradiance (radiant flux intercepted per unit area: unit = w/m^2) or an illuminance (luminous flux intercepted per unit area, unit lux, 10.76 lux = 1 foot candle). It is to note that an irradiance measurement is not spectrally defined, whereas, an illuminance measurement indicates the level of visible light as the human eye would see it. Artificial light is provided by cool, white, fluorescent tubes and/or incandescent bulbs of different classes of artificial light sources. Fluorescent sources have been used almost exclusively for plant tissue culture because they are more efficient at producing broad band visible light than incandescent bulbs, and are available in low output wattage than lamps.

Light intensity can be measured by photometer or lux meter (Fig. 2.30). A photometer consists of photoelectric cell and a micro–ammeter. Photoelectric cell is sensitive for light and converts light into current. Microammeter shows reading due to this current and its needle moves. Nowadays digital read out is given by appropriately converting the current into digital signal. Light intensity is proportional to the current generated in the photoelectric cell by falling light. The unit of illumination is called as 'lux' and the scale of ammeter is calibrated in lux.

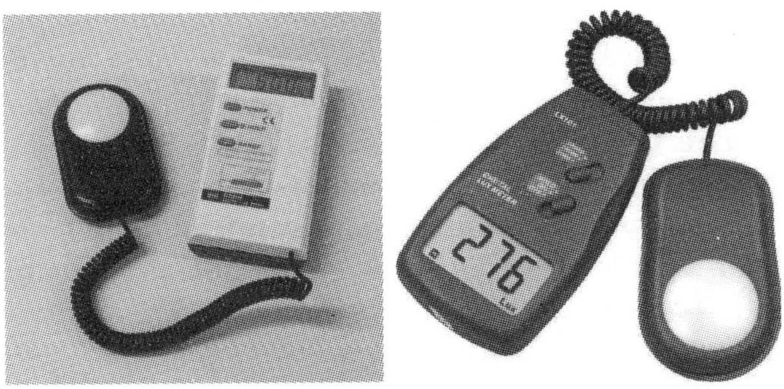

Fig. 2.30: Photometer or luxmeter

Usually 2000 to 2500 lux is provided to the cultures maintained in light in the culture room. The instrument is a sensitive tool and should be handled with care. It should not be exposed to the sunlight.

2.22. LABORATORY SAFETY

Laboratories contain hazards of various kinds students are required to wear trousers and a lab load inside the laboratory without instructor students should not work in the laboratory. Some of the chemicals used are toxic, mutagenic or teratogenic, hence students should take maximum precautions while handling these hazardous chemicals.

Chapter 3

Methods of Sterilization

Plant tissue culture work requires strict maintenance of asepsis (germ free environment) in all the operations. Particularly in commercial production units, the contamination of one batch of the cultures may result in heavy financial losses or even loss of a culture strain. Therefore, strict control measures are enforced to maintain the entry of the personnel and living materials. The basic rules and practices of asepsis are followed in all the tissue culture laboratories.

3.1 CLEANLINESS OF LABORATORY

Plant tissue culture media are rich in nutrients and very suitable for the growth of microbes also. These microorganisms grow faster, consume nutrients rapidly and suppress the growth of plant tissues (by over growth). This saprophytic unwanted growth of microbes is termed as contamination. There are many ways by which these microbes suppress the growth of plant cells and tissue, *e.g.*, release of enzymes and toxins (which inhibit the growth of plant cells), selective absorption of specific nutrients, and some tissue infection. To achieve success in cultivation of higher plant parts, it is essential to exclude these contaminating microorganisms and hence aseptic techniques must be employed to save cultures. This has led to the development of 'clean area concept'. All the area, tools and working places should be free from microbes to get rid of problems of contamination. Therefore, regular cleaning and sanitation of the laboratory is very important (Fig. 3.1).

Fig. 3.1: Cleanliness of laboratory

Following care should be taken in a tissue culture laboratory:

1. Minimize the air current in the working area to prevent spores of contaminating microorganisms to move in along with the air currents over the sterile areas. At least, fan should not be used in the inoculation room. Preferably, all the places should be air conditioned.
2. Store the prepared media, nutrients and tools in cabinets properly.
3. Use separate areas for cleaning and washing of glass vessels and for the preparation of medium.

Fig. 3.2: Glassware racks

Inoculation room and culture room should be isolated and movement of personnel should be restricted. Use of positive air pressure, air curtain and removal of shoes outside the culture room, minimize the problem of contamination. Use of apron, cap and face mask help in preventing the contamination. Inoculation room and corridors may be provided with germicidal ultra–violet (UV) tube lights. UV tubes are also fixed in Laminar air flow bench (Fig. 3.3) placed in inoculation chamber. All these UV tube lights should be used frequently before inoculations.

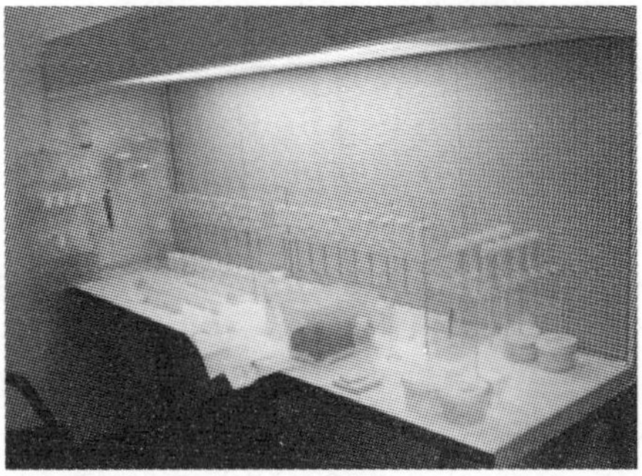

Fig. 3.3: Laminar air flow hood

3.2 STERILIZATION OF TOOLS AND MEDIA

All the manipulations of plant tissue culture methods are carried out in aseptic conditions. All the materials used are therefore, free from microbes or sterilized. Various methods of sterilization are used depending upon the material and types of sterilization. Essentially, method for sterilization is different for living materials than tools and glass wares. Nowadays many items are available as ready to use, pre–sterilized (by gamma radiation) and disposable. Use of such disposable materials has facilitated the work. In laboratory, sterilization is carried out mainly in three ways.

1. Dry heat sterilization.
2. Wet heat sterilization.
3. Filter sterilization.

3.2.1 Dry Heat Sterilization

Glassware, metal tools and other articles, which can tolerate high temperature are put in containers or wrapped in paper or thick aluminium foil and placed in hot air oven (Fig. 3.4) for a period of 1–3 hours at a temperature of 140–160°C for dry heat sterilization. Media and plasticwares can not be sterilized by this method.

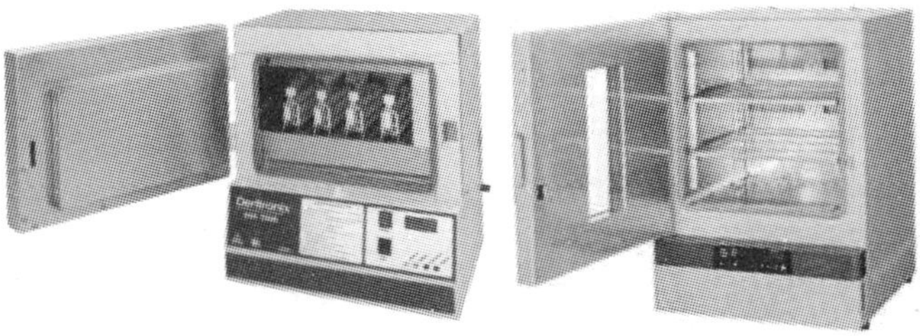

Fig. 3.4: Dry heat sterilizers (hot air ovens)

3.2.2 Wet Heat Sterilization

It is the most popular method of sterilization. Both surgical items and media are autoclaved at 121°C with a pressure of 15 psi (pounds per square inch) or 1.02 kg/cm^2 for 15 min. Modern autoclaves are capable of providing saturated steam treatment ranging from 70–132°C, at a pressure of up to 25 psi. All the vessels containing medium should be placed vertically and should not be filled more than 40% to their total capacity. There is a relationship between amount of medium to be sterilized and duration of autoclaving (Table 3.1). Larger volumes require longer duration for autoclaving at 121°C.

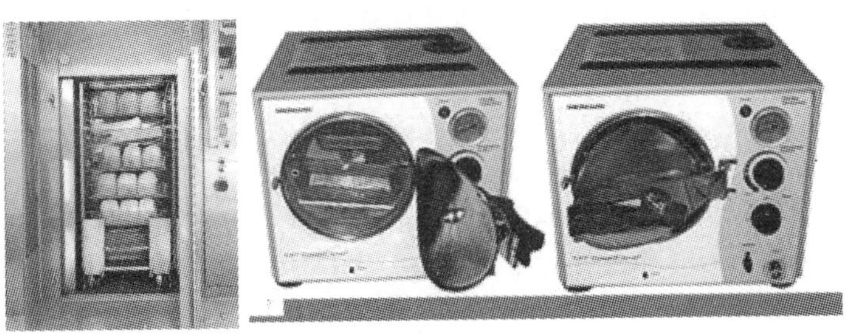

Fig. 3.5: Autoclave horizontal

Table 3.1: Sterilization period for various volumes of medium

Volume (ml)	*Duration at 121°C (min)*
20–25	15
75	20
250–500	25
1000	30
1500	35
2000	40

3.2.3 Filter Sterilization

Filter sterilization or cold sterilization is used when a solution or medium cannot be sterilized by autoclaving. It is the property of the filter (porosity 0.22 to 0.45 μm) to retain all the microorganisms and make the solution free from microbes. This exclusion of microorganisms makes the solution sterilized without heating or autoclaving. Only liquids can be sterilized by this method and not the plant materials. This method is commonly used to sterilize thermo labile or heat sensitive chemicals like plant growth regulators (IAA, GA_3, Zeatine, Abscisic acid), some biochemical, enzymes for protoplasts isolation, antibiotics or nutrient solution for specific experiment in order to avoid decomposition during autoclaving. After filter sterilization, these solutions are added to the medium. In case of static medium, compound is added before solidification (cooling) after autoclaving.

Glass and membrane filters (nitro–cellulose membranes) are commercially available. Glass filters (sintered glass filters, Borosil), available in different grades (G–1 to 5), is used for bacteriological purposes. These filters have a porous glass disc for filtration of liquids and gases as a filter media which is non–corrosive and reusable. The grades are classified by maximum pore size which is obtained by measuring the pressure at which the first air bubble breaks away from filter under certain conditions. The pressure differential is then used to calculate the equivalent capillary diameters in microns. The desired *pore* size is obtained by suitably controlling the *grain* size, firing time and temperature and the thickness of the disc. These filters are resistant to heat, solvents and detergents.

Membrane type filters (of different diameter and porosity ranging from 0.22 to 0.45 µm) are available as disposable or reusable filters, or as filter holders containing a filter disc (Fig. 3.7). These filters can be connected through tubing or syringe to pass solutions or gases.

Filters are wrapped in aluminium foil or paper, autoclaved at 121°C for 15 min and then taken to laminar air flow bench. Solution is placed in funnel (glass type filter) or syringe (membrane type filter) and solution is allowed to pass through filter under suction (glass) or pressure (membrane). A membrane filter or glass tube, filled with cotton may be placed between vacuum flask and suction pump to avoid entry of air from the outside during operation. Sterilized filtrate is collected in flask and dispensed in the liquid or molten medium with the help of sterilized pipette lined with cotton plug. Volume added to the medium is adjusted to arrive at a final concentration. Membrane type filters are also used as inlets and outlets of various types including air, in a bioreactor system and to draw samples.

Fig. 3.6: Vertical autoclave

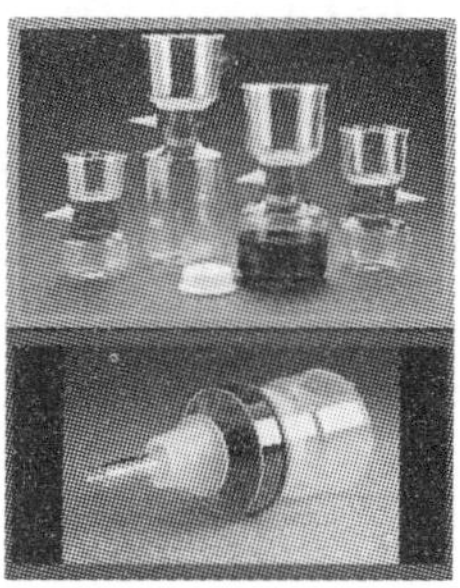

Fig. 3.7: Glass and membrane filters

Besides wet and dry heat treatments, radiations and other methods such as filtration, gas and chemical treatments are also available for direct sterilization of material. Gas treatments are rarely used in the laboratory. Ultraviolet light treatment of working surfaces and sterile rooms are used in the labs, while gamma radiation is used in the industry for the preparation of pre–sterilized disposable plasticwares.

3.2.4 Sterilization of Tools

Surgical blades and scalpels are not sterilized by dry heat because the high temperature makes the cutting edge dull. Such articles including spatula and forceps are usually immersed in 80% ethyl alcohol until required, and sterilized during use by frequent immersion in alcohol and flaming. Nowadays, bead sterilizer (based on principle of dry heat) is available for sterilizing such tools. After an initial sterilization time of 30 minutes during which articles attains temperature of about 250°C, total sterilization by destruction of all microorganisms is ensured within seconds. This unit can conveniently be placed on laminar air–flow bench itself.

Fumigation (with formaldehyde, sulphur etc.) is sometimes used to clear the laboratory or working area from microorganisms. Culture rooms are fumigated only after removal of cultures. Generally, wiping of working places with 80% alcohol is sufficient for regular use.

3.2.5 Mode of Action of Disinfectants

The various metallic ions can be arranged in a series of describing antibacterial activity. Hg^{2+} and Ag^{+}, at the top of the list, are effective at less than 1 ppm, because of their high affinity for–SH (sulfhydryl) group. Bacteria are killed by Ag^{+} containing 10^5–10^7 Ag^{+} ions per cell. The concentration required for killing is therefore markedly affected by the inoculum size. The antibacterial action of Hg^{2+} can be readily reversed by sulfhydryl compounds.

Chlorine was the antiseptic introduced (as chlorinated lime) by O.W. Holmes in Boston in 1835, and by Semmelweis in Vienna in 1847, to prevent transmission of puerperal sepsis by the physician's hands. Chlorine combines with water to form hypochlorous (HOCl) a strong oxidizing agent:

$$Cl_2 + H_2O = HCl + HOCl$$

$$Cl_2 + 2NaOH = NaCl + NaOCl + H_2O$$

Hypochlorite solutions (200 ppm Cl_2) are used to sterilize clean surfaces in the food and the dairy industries and in restaurants, and

Cl_2 gas added at 1–3 ppm is widely used to disinfect water supplies and swimming pools.

3.2.6 Sterilization of Explant

All explants have to be sterilized before transferring them on to the medium (Fig. 3.8). A suitable sized explant can be sterilized by anyone of the following solutions:

1. 1–4% solution of sodium hypochlorite (commercial bleach).
2. 7% saturated solution of calcium hypochlorite.
3. 1% solution of bromine water.
4. 7% ethyl alcohol.
5. 0.1–0.2% mercuric chloride.
6. 10% hydrogen peroxide solution.
7. 1% silver nitrate solution.

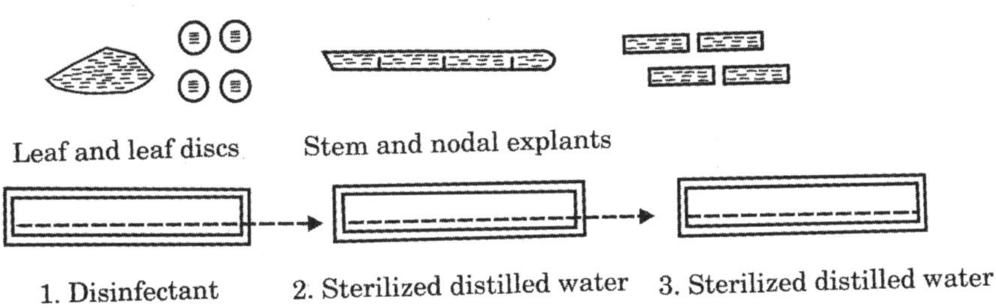

Fig. 3.8: Method of sterilization of explants

The type, concentration and duration of exposure of a particular sterilant to be used depend on the plant material under use, and are determined empirically. In many cases 2 drops of liquid detergent or wetting agent (Teepol, Tween–80) are incorporated into the sterilizing solution to enhance the penetration and effectiveness of the sterilizing agent, Material should be thoroughly washed several times with sterilized distilled water to ensure proper removal of sterilant. All the sterilized materials and plant parts should not be touched directly by hand and sterilized forceps etc should be used. It is obvious that hands and bench are made aseptic by ethanol (80%) before start of work.

Currently, use of toxic heavy metals like mercury is discouraged because it requires proper disposal facility for used chemicals. Therefore, calcium hypochlorite solution is commonly used and it is quite effective. For very specific cases underground plant materials (like roots, tubers and rhizomes), pretreatment with antibiotic is employed. But it is not commonly required for all the materials. These chemicals may be detrimental to the growth of tissues during subsequent growth period. Plant tissue culture work requires strict control on contamination in all the areas of work to ensure continuous healthy growth of the cultures. All the contaminated cultures should be destroyed by autoclaving and then cleaned.

Chapter 4

Culture Medium and its Preparation

4.1 INTRODUCTION

The present day nutrient media that meet the nutritional requirement for plant cell and tissue cultures, were developed during 1930–50. Basically all nutrient media contained the same nutrients as required by whole plant. The discovery of plant growth regulators during that period and knowledge about nutrition of plants, particularly micro–elements and vitamins, contributed significantly in the development of nutrient media. The notable being, the discovery of Indole-3-acetic acid (IAA) by Went, F.W. (1932) and kinetin by Skoog and co–workers (1955). This was followed by demonstration of totipotency and regulation of morphogenesis through a combination of auxin and cytokinin in tobacco by Skoog and Miler in 1957.

The media used by earlier workers were based on Knop's solution. Subsequently media developed by White (1943) and Heller (1953) were used. Murashige and Skoog's medium (Murashige and Skoog, 1962) is a land mark in plant tissue culture research. It is the most frequently used medium for a wide range of higher plants. Based on its composition, other media were evolved to meet the diverse experimental and species specific requirements. These include Linsmaier and Skoog (1965), B_5 medium of Gamborg *et al.* (1968), Schenk and Hildebrandt (1972) medium, Nitsch and Nitsch

(1969) medium, and woody plant medium (Llyod and McCown, 1980).

The methodology of plant tissue culture has advanced to the stage where tissues from virtually any plant species can be cultured *in vitro* successfully. The successful plant tissue culture depends upon the choice of nutrient medium. The cells of most plant species can be grown on completely defined media. All the media consist of mineral salts, a carbon source (generally sucrose), vitamins and growth regulators. The MS medium designed for tobacco is now widely used for various species, in callus and cell cultures.

4.2 MEDIUM COMPOSITION

The nutrient medium for most plant tissue cultures is comprised of five groups of ingredients–inorganic nutrients, carbon source, vitamins, growth regulators and organic supplements (Tables 4.1 to 4.4).

Table 4.1: Common salts used in various media and their molecular weights

Mineral salts	*Molecular weight*
H_3BO_3	61.84
$CaCl_2.2H_2O$	147.02
$Ca(NO_3)_2$ $4H_2O$	236.15
$CoCl_2.6H_2O$	237.93
$CuSO_4.5H_2O$	249.68
EDTA, Sodium ferric salt (13% Fe)	366.85
$FeSO_4.7H_2O$	278.00
KCl	74.56
KH_2PO_4	136.09
KI	166.01
KNO_3	101.10
KOH	56.10
$K_2.SO_4$	174.10
$MgSO_4.7H_2O$	246.50
$MnSO_4.H_2O$ (Note Conversion)	169.10
$MnSO_4.4H_2O$ (Note Conversion)	223.09
NaCl	58.44
$Na_2.EDTA.2H_2O$	372.20

Table 4.1: *(Contd...)*

Table 4.1: (*Contd...*)

Mineral salts	*Molecular weight*
$NaH_2PO_4.2H_2O$	137.98
$NaOH$	40.01
$Na_2MoO_4.2H_2O$	241.95
Na_2SO_4	142.06
NH_4Cl	53.49
NH_4NO_3	80.09
$NH_4H_2PO_4$	115.13
$(NH_4)_2SO_4$	132.14
$ZnSO_4.7H_2O$	287.55

Table 4.2: Vitamins used in different media and their molecular weights

Vitamins	*Molecular weight*
p–Aminobenzoic acid	137.13
Ascorbic acid	176.12
Biotin	244.30
Choline chloride	139.63
Folic acid	441.40
Myo–Inositol	180.16
Nicotinamide (niacinamide)	122.12
Nicotinic acid (niacin)	123.11
Pantothenate, calcium salt	476.53
Pyridoxine hydrochloride	205.64
Riboflavin	576.40
Thiamine hydrochloride	337.28
Vitamin A (retinol)	286.44
Vitamin B_{12}	1355.40

The nutrient medium has to supply all the essential mineral ions required for growth and development. In many cases (as the biosynthetic capability of cells cultured *in vitro* may not replicate to that of the parent plant), it must also supply additional organic supplements such as amino acids and vitamins. Many plant cell cultures, as they are not photosynthetic, also require the addition of a fixed carbon source in the form of a sugar (most often sucrose). One other vital component that must also be supplied is water, the principal biological solvent. Physical factors, such as temperature, pH, the gaseous environment, light (quality and duration) and osmotic pressure, also have to be maintained within acceptable limits.

4.3.1 Plant Cell Culture Media

Culture media used for the *in vitro* cultivation of plant cells are composed of three basic components:

(1) Essential elements, mineral ions, supplied as a complex mixture of salts

(2) An organic supplement supplying vitamins and/or amino acids; and

(3) A source of fixed carbon; usually supplied as the sugar (sucrose).

Table 4.3: Elements required for plant nutrition and their physiological function. These elements have to be supplied by the culture medium in order to support the growth of healthy cultures *in vitro*.

Elements	***Functions***
Nitrogen	Component of proteins, nucleic acids and some coenzymes Element required in greatest amount.
Potassium	Regulates osmotic potential, principal inorganic cation
Calcium	Cell wall synthesis, membrane function, cell signaling
Magnesium	Enzyme cofactor, component of chlorophyll
Phosphorus	Component of nucleic acids, energy transfer, component of intermediates in respiration and photosynthesis
Sulphur	Component of some amino acids (methionine, cysteine) and some cofactors
Chlorine	Required for photosynthesis
Iron	Enzyme cofactor
Cobalt	Component of some vitamins
Copper	Enzyme cofactor, electron–transfer reactions
Zinc	Enzyme cofactor, chlorophyll biosynthesis
Molybdenum	Enzyme cofactor, component of nitrate reductase

4.3.2 Media Components

Macro elements

Elements required in large amounts for plant growth and development are categorized as macro–elements. Nitrogen, phosphorus, potassium, magnesium, calcium and sulphur (and carbon, which is added separately) are usually regarded as macro elements. These elements usually comprise at least 0.1% of the dry weight of plants.

Nitrogen is most commonly supplied as a mixture of nitrate ions (from KNO_3) and ammonium ions (from NH_4NO_3). Theoretically, there is an advantage in supplying nitrogen in the form of ammonium ions, as nitrogen must be in the reduced form to be incorporated into macromolecules. Nitrate ions therefore need to be reduced before incorporation. However, at high concentrations, ammonium ions can be toxic to plant cell cultures and uptake of ammonium ions from the medium causes acidification of the medium. In order to use ammonium ions as the sole nitrogen source, the medium needs to be buffered. High concentrations of ammonium ions can also cause culture problems by increasing the frequency of vitrification (the culture appears pale and 'glassy' and is usually unsuitable for further culture). Using a mixture of nitrate and ammonium ions has the advantage of weakly buffering the medium as the uptake of nitrate ions causes OH^- ions to be excreted.

Phosphorus is usually supplied as the phosphate ion of ammonium, sodium or potassium salts. High concentrations of phosphate can lead to the precipitation of medium elements as insoluble phosphates.

Micro elements

These elements are required in trace amounts for plant growth and development, and have many diverse roles. Manganese, iodine, copper, cobalt, boron, molybdenum, iron and zinc usually comprise the micro elements, although other elements such as nickel and aluminium are frequently found in traces in some formulations.

Iron is usually added as iron sulphate, although iron citrate can also be used. Ethylene diamine tetra acetic acid (EDTA) is used in conjunction with the iron sulphate. The EDTA complexes with the iron is used so as to allow the slow and continuous release of iron into the medium. Uncomplexed iron can precipitate out of the medium as ferric oxide.

Organic supplements

Only two vitamins, thiamine (vitamin B_1) and myo inositol (B vitamin) are considered essential for the culture of plant cells *in vitro*. However, other vitamins are often added to plant cell culture media for historical reasons.

Amino acids are also commonly included in the organic supplement. The most frequently used is glycine (arginine, asparagine, aspartic acid, alanine, glutamic acid, glutamine and proline are also used), but in many cases its inclusion is not essential. Amino acids provide a source of reduced nitrogen and, like ammonium ions, uptake causes acidification of the medium. Casein hydrolysate can be used as a relatively cheap source of a mix amino acids.

Carbon source

Sucrose is cheap, easily available, readily assimilated and relatively stable and is therefore the most commonly used carbon source. Other carbohydrates (such as glucose, maltose, galactose and sorbitol) can also be used (see Chapter 3), and in specialized circumstances may prove superior to sucrose.

Gelling agents

Media for plant cell culture *in vitro* can be used in either liquid or 'solid' forms, depending on the type of culture being grown. For any culture types that require the plant cells or tissues to be grown on the surface of the medium, it must be solidified (more correctly termed 'gelled'). Agar, produced from seaweed, is the most common type of gelling agent, and is ideal for routine applications. However, because it is a natural product, the agar quality can vary from supplier to supplier and from batch to batch. For more specialized applications (*e.g.*, microspore culture), a range of purer (and in some cases, considerably more expensive) gelling agents are available. Purified agar or agarose can also be used as a gelling agent.

4.3.3 Plant Growth Regulators

We have already briefly considered the concepts of plasticity and totipotency. The essential point as far as plant cell culture is concerned is that, due to this plasticity and totipotency, specific media formulations can be used to direct the development of plant cells in culture. Plant growth regulators are the critical media components in determining the developmental pathway of the plant cells. The plant growth regulators used most commonly are plant hormones or their synthetic analogues.

4.3.4 Classes of Plant Growth Regulators

There are five main classes of plant growth regulators used in plant cell and tissue culture, namely:

(1) Auxins;
(2) Cytokinins;
(3) Gibberellins;
(4) Abscisic acid;
(5) Ethylene.

Each class of plant growth regulator has been briefly described as under:

Auxins

Auxins promote both cell division and cell growth. The most important naturally occurring auxin is IAA (indole–3–acetic acid), but its use in plant cell culture media is limited because it is unstable to both heat and light. Occasionally, amino acid conjugates of IAA (such as indole–acetyl–L–alanine and indole–acetyl–L–glycine), which are more stable, are used to partially alleviate the problems associated with the use of IAA. It is more common, though, to use stable chemical analogues of IAA as a source of auxin in plant cell culture media. 2,4–Dichlorophenoxyacetic acid (2,4–D) is the most commonly used auxin and is extremely effective in most circumstances. Other auxins are available (see Table 4.4), and some of them may be more effective or 'potent' than, 2,4–D in some instances.

Cytokinins

Cytokinins promote cell division. Naturally occurring cytokinins are a large group of structurally related (purine derivatives) compounds.

Table 4.4: Commonly used auxins, their abbreviations and chemical names

Abbreviation name	***Chemical name***
2,4 – D	2, 4–dichlorophenoxyacetic acid
2, 4, 5–T	2, 4, 5–trichlorophenoxyacetic acid
Dicamba	2–methoxy–3, 6–dichlorobenzoic acid
IAA	Indole–3–acetic acid
IBA	Indole–3–butyric acid
MCPA	2–methyl–4–chlorophenoxyacetic acid
NAA	I–naphthyl acetic acid
NOA	2–naphthyloxyacetic acid
Picloram	4–arnino–2, 5, 6–trichloropicolinic acid

Of the naturally occurring cytokinins, two have some use in plant tissue culture media (see Table 4.5). These are zeatin and 2iP (2–isopentyl adenine). Their use is not widespread as they are expensive (particularly zeatin) and relatively unstable. The synthetic analogues, kinetin and BAP (benzylaminopurine), are therefore used more frequently. Non–purine–based chemicals, such as substituted phenylurea, are also used as cytokinins in plant cell culture media. These substituted phenylurea can also substitute for auxin in some culture systems.

Table 4.5: Commonly used cytokinins, their abbreviations and chemical names

Abbreviation name	***Chemical name***
BAP[a]	6–benzylaminopurine
2iP (IPA)[b]	[N^6–(2–isopentyl) adenine]
Kinetin[a]	6–furfurylaminopurine
Thidiazuron[c]	1–phenyl–3–(1, 2, 3–thiadiazol–5–yl) urea
Zeatin[b]	4–hydroxy–3–methyl–trans–2–butenylamino-purine

[a]Synthetic analogues, [b]Naturally occurring cytokinins, [c]A substituted phenylurea–type cytokinin.

Gibberellins

There are numerous, naturally occurring, structurally related compounds termed 'gibberellins'. They are involved in regulating cell elongation, and are agronomically important in determining plant height and fruit–set. Only a few of the gibberellins are used in plant tissue culture media, GA_3 being the most common.

Abscisic acid

Abscisic acid (ABA) inhibits cell division. It is most commonly used in plant tissue culture to promote distinct developmental pathways such as somatic embryogenesis.

Ethylene

Ethylene is a gaseous, naturally occurring, plant growth regulator most commonly associated with controlling fruit ripening in climacteric fruits, and its use in plant tissue culture is not widespread. Some plant cell cultures produce ethylene which, if it builds up sufficiently, can inhibit the growth and development of the culture. The type of culture vessel used and its means of closure affect the gaseous exchange between the culture vessel and the outside atmosphere and thus the levels of ethylene present in the culture.

4.4 PLANT GROWTH REGULATORS AND TISSUE CULTURE

Generalizations about plant growth regulators and their use in plant cell culture media have been developed from initial observations made in the 1950s. There is, however, some considerable difficulty in predicting the effects of plant growth regulators: this is batches of the great differences in culture response between species, cultivars and even plants of the same cultivars grown under different conditions. However, some principles do hold true and have become the paradigm on which most plant tissue culture regimes are based.

Auxin to cytokinin high ratios promote root development, low ratios promote shoot development. Intermediate ratios promote continued growth of the callus without differentiation. Auxins and cytokinins are the most widely used plant growth regulators in plant tissue culture and are usually used together, the ratio of the auxin to the cytokinin determining the type of culture established or regenerated (see Fig. 4.1). A high auxin to cytokinin ratio generally favours root formation, whereas a high cytokinin to auxin ratio favours shoot formation. An intermediate ratio favours callus production.

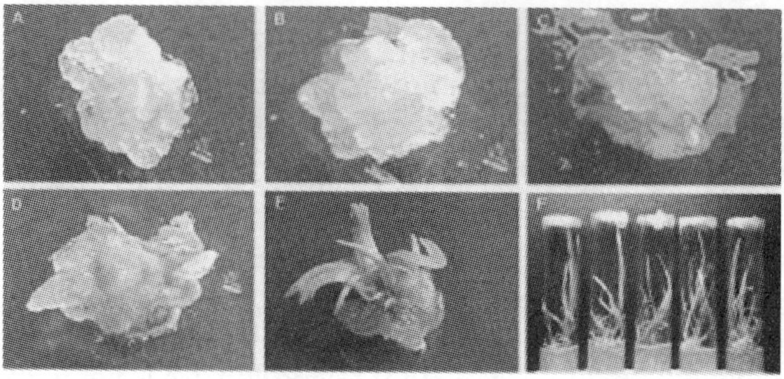

Fig. 4.1: The effect of different ratios of auxin to cytokinin on the growth and morphogenesis of callus

4.5 MEDIUM PREPARATION

Glass distilled water and chemicals of highest purity should be used. A convenient approach to prepare a medium is to have stock solutions of all the nutrients in a 10x or 50x concentration. Medium is prepared by suitably diluting the appropriate amount of stock solutions for desired volume of the medium. Stock solutions are stored in a refrigerator. It may be advantageous to have separate stock solutions of calcium salt and potassium iodide. NAA, 2, 4–D and similar compounds are dissolved in small amount of ethanol and made to desired volume with water. The cytokinins are dissolved in a small amount of 0.5 N HCI with slight heat and then made to volume with water. All the ingredients are mixed, sugar is added, pH is adjusted to 5.8–6.0 and medium is poured in the culture vessels. All the vessels are plugged with non–absorbent cotton plugs, covered with aluminium foil, and autoclaved at 121°C for 15 min. All the ingredients should be thoroughly mixed before dispensing in the vessels. Autoclaved media should be stored in the cabinets placed in the inoculation room immediately after sterilization.

Composition of MS and B_5 Media

Salts	***MS***		***B_5***	
A. Macronutrients	mg/l	μM	mg/l	μM
NH_4NO_3	1650	20.6	–	–
KNO_3	1900	18.8	2500	25
$CaCI_2.2H_2O$	440	3.0	150	1.0
$MgSO_4.7H_2O$	370	1.5	250	1.0
KH_2PO_4	170	1.25	–	–
$(NH_4)_2SO_2$	–	–	134	1.0
$NaH_2PO_4.H_2O$	–	–	150	1.1
B. Micronutrients	mg/l	µM	mg/l	µM
KI	0.83	5.0	0.75	4.5
H_3BO_3	6.2	100	3.0	50
$MnSO_4.4H_2O$	22.3	100	–	–
$MnSO_4.H_2O$	–	–	10	60
$ZnSO_4.7H_2O$	8.6	30	2.0	7.0
$Na_2MoO_4.2H_2O$	0.25	1.0	0.25	1.0
$CuSO_4.5H_2O$	0.025	0.1	0.025	0.1
$CoCl_2.6H_2O$	0.025	0.1	0.025	0.1
Na_2EDTA	37.3	100	37.3	100
$FeSO_4.7H_2O$	27.8	100	27.8	100
Sucrose (g)	30,000	–	20,000	–
pH	5.7–6.0	–	5.5–5.8	–
C. Vitamins				
Inositol	100		100	–
Nicotinic acid	0.5		1.0	–
Pyridoxine HCI	0.5		1.0	–
Thiamine HCI	0.1		10.0	–
D. Amino acid				
Glycine	2.0		–	–
E. Plant Growth Regulators				
NAA/IAA/2,4–D	0.01–10.0	–	–	–
Kinetin	0.04–10.0	–	–	–

Table 4.6: Plant growth regulators used in different media and their molecular weights

Plant growth regulators	***Abbreviation***	***Molecular weight***
Abscisic acid	ABA	264.3
Adenine	ADE	135.1
Ancyrnidol	ANC	256.3
N^6–Benzyladenine (6–benzylaminopurine)	BA	225.3
Chlorocholine chloride	CCC	158.1
p–Chlorophenoxyacetic acid	CPA	186.6
Dicamba (3, 6–dichloro–o–anisic acid)	DCA	221.0
2, 4–Dichlorophenoxyacetic acid	2,4–D	221.0
6–(y, y–Dirnethylallyamino) purine (2–isopentenyladenine)	2iP	103.2
Gibberellic acid	Gk	330.0
Indole 3–acetic acid	IAA	175.2
Indole 3–butyric acid	IBA	203.2
Jasmonic acid	JA	210.3
Kinetin (6–furfurylarninopurine)	KIN	215.2
a–Napthaleneacetic acid	NAA	186.2
Picloram (4–amino–3, 5, 6–trichloropicolinic acid)	PIC	241.5
Silver nitrate	$AgN0_3$	169.9
Thidiazurone [N–phenyl–N–(1, 2, 3–thiadiazol–5–yl) urea]	TDZ	220.2
2, 4, 5–Trichlorophenoxyacetic acid	2, 4, 5–	255.5
Zeatin	ZEA	219.2

4.5.1 Preparation of Stock Solutions

It is not convenient to weigh and mix all the constituents separately for the preparation of medium. It is time–consuming and a tedious job, Again if 100 ml or 200 ml medium is to be prepared, it is very difficult to weigh some constituents that are used in very small quantity. So it is convenient to prepare the concentrated stock solutions of macrosalts, micro–salts, vitamins, amino acids, hormones etc. All stock solutions should be stored in a refrigerator and should be regularly checked visually for contamination with microorganisms or precipitation of ingredients, Stock solution of vitamins, amino acids and hormones should not be stored for indefinite period and should be kept in a deep freezer. The widely used culture medium was formulated by Murashige and Skoog (commonly called MS medium), so the procedure for the preparation of stock solution of MS medium is given below:–

Constituents	*Amount (mg/L) present in original medium*	*Amount (g) to be taken for stock solution (20X)*	*Final volume of stock (ml)*
NH_3NO_4	1650	33.0	–
KNO_3	1900	38.0	1000
$CaCl_2.2H_2O$	440	8.8	–
KH_2PO_4	170	3.4	–
$MgSO_4.7H_2O$	370	7.4	–

Particularly macro–salts should be made at ten/twenty times of (10x or 20x) their original concentration in the medium while micro–nutrients can be prepared at thousand time (1000x) of their final strength. While making the stock solutions, it is advisable to dissolve each constituent completely before adding another otherwise precipitation of salts may occur.

4.5.1.1 *Stock solution of macro–salts (20X)*

To make 1,000 ml of this stock solution, dissolve chemicals one after another in 800 ml of double distilled water, (DDH_2O) and then make up the final volume. The solution is filtered and can be stored in a refrigerator (10–16°C) until the solution is totally used.

4.5.1.2 *Stock solution of KI (1000X)*

Dissolve 83 mg of KI (0.38 mg/l present in original medium) in 100 ml of double distilled H_2O. Store in refrigerator at 10–16°C.

4.5.1.3 *Stock solution of iron (200X)*

Dissolve at first 745 mg of Na_2 EDTA (37.25 mg/l in original) in 75 ml boiling double distilled water, then add gradually 557 mg of $FeSO_4.7H_2O$ (27.85 mg/l in original). Keep on a magnetic stirrer for at least 1 hr in hot condition until the colour of the solution changes to golden yellow. Finally make the volume to 100 ml and store in refrigerator (5ºC). This solution specifically must be kept in an amber coloured bottle.

To make 100 ml of this stock solution, dissolve the salts sequentially as shown above one by one in 80 ml of DDH_2O. Make up the final volume. Filter and store the solution at approximately 5ºC in a refrigerator.

4.5.1.4 *Stock solution of meso–inositol (500X)*

Dissolve 1 gm meso–inositol (100 mg/l in original) in 20 ml DDH_2O. Store at 0°C for 15 days.

4.5.1.5: *Stock solution of micro–salts (1,000 x)*

Constituents	*Amount (mg/L) present in original medium*	*Amount (X100) to be taken for stock solution (Value expressed in mg)*	*Final volume of stock (ml)*
H_3BO_3	6.2	620	100
$Na_2MoO_4.2H_2O$	0.25	25	–
$CoCl_2.6H_2O$	0.025	2.5	–
$CuSO_4.5H_2O$	0.025	2.5	–
$ZnSO_4.7H_2O$	8.6	860	–
$MnSO_4.4H_2O$	22.3	2230	–

4.5.1.6: *Stock solution of vitamins (1,000 X)*

Constituents	*Amount (mg/L) present in original medium*	*Amount (50X) to be taken for stock solution (value in mg)*	*Final storage volume (ml)*	*Temp-erature (0ºC)*	*Duration of storage (in days)*
Thiamine–HCl	0.1	5	–	–	–
Nicotinicacid	0.5	25	50	0	15
Pyridoxine–HCl	0.5	25	–	–	–

4.5.1.7: *Stock solution of hormones*

Hormones	*Required amount for stock solution (mg)*	*Amount of solvent required to dissolve*	*Amount of water to be added*
Auxins			
2,4–dichloro–phenoxy acetic acid (2,4–D)	5	1 ml absolute ethyl alcohol	9 ml
Indole–3–acetic acid (IAA)	5	do	do
α–naphthalene acetic acid (NAA)	5	do	do
3–indole butyric acid (IBA)	5	do	do
Cytokinins			
Kinetin (6–furfuryl amino purine)	5	1 ml of 1 (N) HCl	do
BAP (6–Benzyl–aminopurine)	5	do	do
Zeatin	5	do	do
2iPA (2, iso–pentenyl–adenine)	5	do	do

Hormones	*Final concent-ration*	*Storage tempera-ture (0°C)*	*Duration of storage (days)*
Auxins			
2, 4–dichloro–phenoxy acetic acid (2,4–D)	0.5 mg/ml	0	7
Indole–3–acetic acid (IAA)	do	do	do
α–naphthalene acetic acid (NAA)	do	do	do
3–indole butyric acid (IBA)	do	do	do
Cytokinins			
Kinetin (6–furfuryl amino purine)	do	do	do
BAP (6–Benzyl–aminopurine)	do	do	do
Zeatin	do	do	do
2iPA (2, iso–pentenyl–adenine)	do	do	do

4.5.1.8 *Stock solution of glycine (1,000x)*

Dissolve 40 mg glycine in 20 ml of double distilled water. Store at 0°C and use within 15 days.

4.5.1.9 *Stock solution of hormones*

The stock solutions of hormones can be used in any medium in desired combinations and concentrations. Auxins and cytokinins are not soluble in water. So they are first dissolved in water miscible solvents and then water is added to get the final volume.

Auxins and cytokinins are available either in powder or in crystal forms and the stock solutions of commonly used auxins and cytokinins should be prepared as described above.

4.6 TECHNIQUES IN PLANT TISSUE CULTURE

Several techniques have been adopted for *in vitro* plant cell, tissue and organ culture. Among them some are general techniques that are essentially followed in all experiments. There are also some specific techniques that are adopted and designed according to the objectives of the experiment.

4.6.1 Preparation of Culture Medium

Principle

In vivo plant cells, tissues and organs get their appropriate nutrient and growth requirements from the intact plant body for their organised growth and development. Isolated cell, tissues and organs also need nutrients for their *in vitro* growth and development. So, nutrients are supplied artificially according to the medium formulated by several workers. The main objective of medium preparation is to culture the cell, tissue and organ *in vitro*.

Procedure

Media should be prepared with care and the following procedure is recommended.

To make 1 litre of MS medium–

(i) Dissolve 30g cane sugar in 200 ml DDH_2O. Mix 1–2 g activated charcoal and filter through filter paper, set inside the Buchner funnel fitted on a special official flask with small side arm attachment. Filtering is done by using a suction pump.

(ii) Take DDH_2O in another flask and add in sequence the appropriate amount of stock solutions.

Preparation of MS medium from stock solutions	
Stock solution of macrosalts	50 ml
Stock solution of macrosalts	1 ml
Stock solution of Kl	1 ml
Stock solution of Fe–EDTA	5 ml
Stock solution of vitamins	1 ml
Stock solution of Glycine	1 ml
Stock solution of meso–inositol	2 ml

Desired concentration of auxin and/or cytokinin are added from stock solution according to the formula–

$$\frac{\text{Desired concentration}}{\text{Stock concentration}} = \text{Amount (ml) of stock solution to be taken for one litre medium}$$

If the quantity of the medium is less than one litre, then hormones are added using another formula–

$$\frac{\text{Required concentration} \times \text{Volume of Medium}}{\text{Stock concentration} \times 1{,}000} = \text{Amount (ml) of stock solution to be added}$$

(iii) Pour filtered sucrose solution and salts, vitamins, amino acids, hormones solution mixture into a one litre measuring cylinder. Make the final volume to one litre with DDH_2O. Shake well to mix up uniformly.

(iv) Adjust the pH of the liquid medium 5.8–6.0 using 0.1 (N) HCl or 0.1 (N) NaOH with the help of a pH meter.

(v) Add 5% to 8% agar to the liquid medium to make solid medium. Heat to 60°C to dissolve the agar completely.

(vi) Dispense the culture medium into culture tube (20 ml/tube) or wide mouth conical flask (25–40 ml/flask). Insert non–

absorbent cotton plug wrapped with gauge cloth. Cover the plug with the help of brown paper and rubber band.

(vii) Medium is finally sterilized by autoclaving.

4.6.2 Sterilization of Media

Principle

The culture medium, especially when it contains sugar, will also support the growth of micro–organisms like bacteria, fungi etc. So if they come in contact with medium either in cellular form or in spore form, the micro–organisms grow faster than the plant tissues due to their brief life cycle and contaminate the culture. The micro–organisms may come from glass vessels, instruments, nutrient medium used for culture and even from plant material itself. Therefore, the surface of plant tissue and all non–living articles including nutrient medium should be sterilized.

Procedure

(i) *Sterilization of non–living articles*: The routine sterilization of non–living articles such as nutrient medium, glass goods, distilled water, surgical items (wrapped with brown paper) is carried out by autoclaving under steam pressure of 15 lb/ in^2 (at 120ºC temperature) for 15 minutes. While autocalving screw capped glass vessels, care should be taken to ensure that the caps are not closed too tightly so that gases can expand without the risk of explosion.

Thermolabile compounds are often added in the medium and such medium is sterilized either at room temperature or in cold by passing through bacterial filter.

An alternative method of sterilizing glass vessels and surgical instruments is heating in an oven at 150ºC for 3–4 hrs.

(ii) *Sterilization of plant material*: Plant material which is to be cultured, should be surface sterilized to remove the surface borne microorganisms. This procedure is done in front of a laminar air flow or inside the inoculation chamber before the plant material is inoculated onto the culture medium (Fig 4.2).

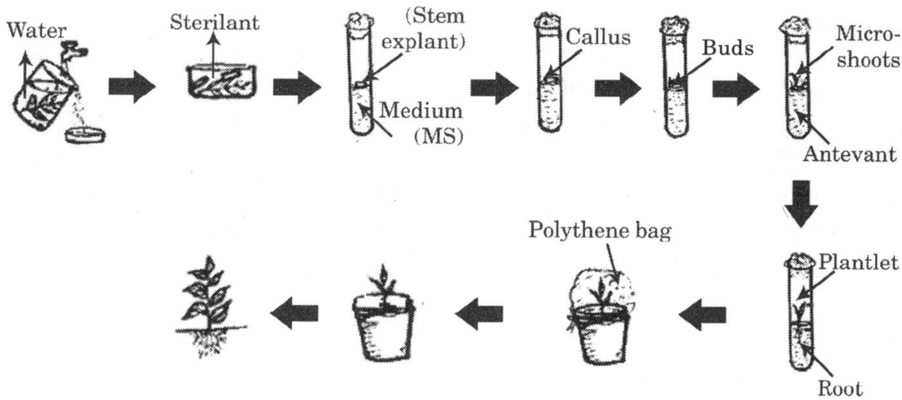

Fig. 4.2: Micropropagation of Plant

(a) Thoroughly washed plant material or explant in tap water is immersed in 5% v/v solution of liquid detergent, such as 'Teepol' for 10–15 minutes. Then wash the material thoroughly in tap water and finally in distilled water. This step can be done in the general laboratory. Further steps of sterilization are done in a laminar air flow or the pre–sterilized inoculation chamber.

(b) Dip the explants in 70% ethyl alcohol for 60 seconds.

(c) Immediately transfer the material into an autoclaved jaw bottle and pour 0.1% mercuric chloride ($HgCl_2$) or 5–10% Sodium hypochlorite (v/v) solution. Keep them for 15 minutes. During that period, the bottle is frequently shaken so that surfaces of plant material come in direct contact with the sterilant.

(d) After 10–15 minutes, decant the solution and wash the explants thoroughly with several changes of autoclaved distilled water to remove all traces of sterilant.

(e) Then the explants are ready for inoculation on nutrient media.

4.7 RAISING OF ASEPTIC PLANTS FROM SEED

Principle

Surface sterilization of explants (plant tissue) may cause some deleterious effect because most of the sterilants are toxic chemicals.

Seeds can more or less resist such deleterious effect due to the presence of its seed coat. So to avoid the contamination of plant tissue, seeds are surface sterilized and are cultured on simple basal nutrient medium. Seeds in culture germinate and give rise to an aseptic seedling. Explants from such seedlings grown under aseptic and controlled conditions are the most suitable material for culture and need no further surface sterilization. Various steps involved in raising aseptic plants (cultures) from seed has been illustrated in Fig. 4.3.

Procedure

(1) Wash the dry seeds thoroughly with tap water.

(2) Dip the seeds in 5% Teepol solution (v/v) for 10–15 minutes. Decant the Teepol solution and wash the seeds again with tap water and finally with distilled water.

(3) Rinse the seeds with 70% ethyl alcohol for 1 minute.

(4) In a laminar air flow, transfer the seeds into an autoclaved bottle and pour 0.1% $HgCl_2$ solution (w/v) so that seeds are immersed. Leave for 10–15 minutes. Stir the bottle frequently.

(5) Decant the sterilant and wash 3–4 times with autoclaved distilled water.

(6) Transfer the seeds from bottle to autoclaved petridish with the help of sterile forceps.

(7) Open the closure of the culture vessel containing the nutrient medium. Flame the neck of the culture vessel and in quick

Fig. 4.3: Raising of aseptic plants from seeds.

succession transfer a few seeds on to the medium. Replace the closure.

(8) Incubate the seeds in continuous dark either at room temperature or at 25–28°C.

4.8 ASEPTIC CONDITION

Precautions must be taken to prevent the entry of any microorganism at the time of transferring the surface sterilized explants on the nutrient medium (inoculation) using the sterilized instruments. For this reason, manipulation and transfer should ideally be carried out under aseptic condition. Starting from surface sterilization to inoculation, all operations should be done aseptically under laminar air flow cabinet.

Procedure

A typical procedure of aseptic technique is given below–

(1) Put all the sterilized articles (media, instruments, glass goods etc.) for inoculation on the glass racks of the inoculation chamber. Alternatively, if laminar air flow is available, keep all articles on the table of air flow cabinet. Laminar air flow blows bacteria free air over the working surface.

(2) Put on the switch of UV lamps of inoculation chamber for one hour before the work is started. In case of laminar air flow, the power switch is put on and the air flow is allowed for at least 15 minutes before work.

(3) Put off the UV lamp before entering inside the inoculation chamber. Do not put off laminar air flow. The working glass table, top of the inoculation chamber or the table of laminar air flow is swabbed with alcohol before starting the work.

(4) Wear a clean apron and use a mask. Clean the hands with alcohol and dry it.

(5) Pour alcohol in a clean coupling jar and dip all instruments into it. Lit the spirit lamp. Take the surface sterilized or aseptic plant material in a sterile Petri dish.

(6) Flame the neck of culture tube or flask and in quick succession remove the plug of glass vials. Transfer the tissue

onto the medium and replace the closure. Each time, the instruments are passed through the flame of the spirit lamp.

Precautions

(1) Always keep away the hands moistened with alcohol from the spirit lamp.
(2) Exposure to UV light builds up a high concentration of ozone gas (toxic) inside the closed chamber. It is, therefore, healthy to enter the chamber only 15–30 minutes after switching off the UV lamp.
(3) Do not dip hot instruments in alcohol and don't use hot instrument for cutting or holding the plant material.
(4) Work carefully and try to ensure that media and plant tissues are exposed only during inoculation.
(5) Don't heat the neck of the glass vials excessively.

4.9 INCUBATION OF CULTURE

Principle

High temperatures are likely to lead to dissociation of the culture medium, and tissue damage while at very low temperatures tissue growth is slow. Again some tissues grow well in dark while others need both light and dark condition. Low humidity causes the quick desiccation of culture medium and high humidity is favourable for the contamination of culture medium. Therefore, cultures are incubated in a culture root where light, temperature and humidity are controlled.

Procedure

(1) After inoculating the tissue to the culture medium, cultures are incubated on culture rack at 25–28°C constant temperature in the growth room.
(2) Culture tubes are placed at 30–45°C in inclined position. For this purpose a long worked stick or an empty paper cover of fluorescence tube is placed on the middle of culture rack and lay the plugged end of the culture tube on the support.

(3) Illumination is provided by cool–white fluorescent light placed about 18 inches above the culture to give a light intensity of 4–10 × 10^3 lux for 16 hours.

(4) If light is not necessary, then put off the light and cover the whole rack with a black cloth or sheet of black paper.

Chapter 5

Culture Initiation

5.1 INTRODUCTION

Plant parts are used to initiate and establish *in vitro* cultures. In case of field or pot grown plants, growth is affected by various soil and environmental factors of variable nature. Therefore, it is difficult to decide the factors affecting the growth, differentiation and metabolism of the plant. Plant tissue culture provides an excellent system to study the plant growth and differentiation in totally controlled conditions. Cultures are grown in a completely defined chemical and physical environment. Thus, by changing the factors, one by one, the influence of a particular factor can be determined very easily. The field grown *(in vivo)* plants are autotrophic, *i.e.*, they synthesize their own food by photosynthesis. In contrast to this, sugar is essential in the medium for the growth of *in vitro* culture, *i.e.*, they are unable to synthesize their carbohydrate requirement (even though cultures are grown in light, photosynthesis is almost negligible).

5.2 EXPLANT TYPE

The present knowledge permits the use of any plant part as a source of material to initiate cultures. The plant part used for this purpose is known as an 'Explant'. Nodal and internodal segments of stem, apical and auxiliary buds, leaf, leaf discs, petiole, anther, pollen, flower bud, petal, inflorescence, ovule, ovary, root and even isolated epidermal peel, gland and trichome have been used as an explant. Selection of any plant part depends upon the objective of study to initiate the cultures. In laboratory, Jerusalem artichoke (*Helianthus tuberous*) tubers or carrot roots are used to prepare disc and cylinders to initiate cultures. Underground material should be thoroughly cleaned to remove dust, crevices, scales, appendages etc. to facilitate sterilization.

5.2.1 Preparation of Explants

Plant material should preferably be brought to the laboratory under ice/dry ice to minimize loss in metabolic activity. Material is washed thoroughly with tap water using a liquid detergent (*e.g*. Tween–80). After proper cleaning and removal of adhering soil particles, material is sterilized and used. Some hard seeds (like leguminous and hard seed coat seeds) may require pretreatment (2–10 min), washed thoroughly after acid treatment under running tap water for 1–2 hours. When axillary or apical buds are used, extra leaves are removed and pre–treated with a wetting agent (5–10 min dip in detergent solution) to facilitate penetration of disinfectant. Explants are prepared in suitable size for the purpose of sterilization, inoculation to generate culture. Stem segments consist of single node or multiple nodes, or internodes are used while leaves can be used as whole leaf or leaf discs.

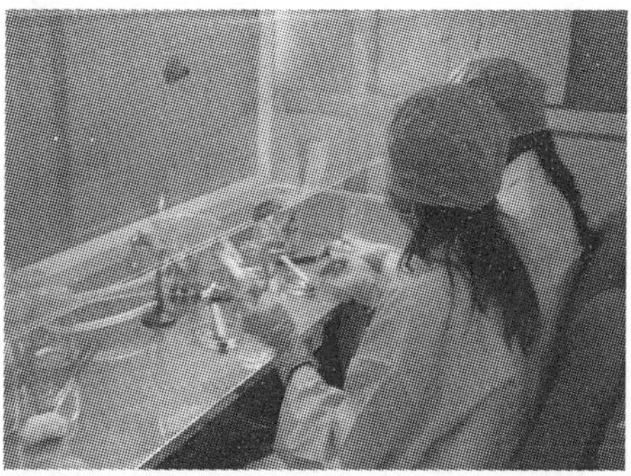

Fig. 5.1: Inoculation of explants

5.2.2 Stem Explants

When stem segments are used to initiate the cultures, the cut ends are sealed with molten wax, and then sterilized with any disinfectant; and washed thoroughly with sterilized distilled water. Sterilized stem pieces are transferred in a pre–sterilized petri dish or sterilized filter paper and ends are removed with the help of scalpel. Then, explants of suitable size consisting of nodes are prepared and transferred on to the medium.

5.2.3 Pollen/Small Seeds

Pollen grains, anthers and small seeds can be sterilized by enclosing them in small filter paper bags (like tea sacs) or enclosing them in a thin sheet of absorbent cotton for ease of handling.

5.2.4 Leaf Disc and Tuber

Leaf discs are prepared after sterilizing the whole leaf using sterilized filter paper pad and a cork borer. In case, leaf is of very large size, pieces or discs can be prepared and then sterilized.

When tuberous materials like Jerusalem artichoke *(Helianthus tuberosus),* sugar beet roots *(Beta vulgaris)* or carrot roots *(Daucus carota)* are used as an explant, large pieces/complete tubers are sterilized and then with the help of scalpel and cork borer, tissue cylinders and discs from cylinders are prepared under aseptic conditions. The explant preparation and sterilization are carried out under laminar air flow bench.

5.3 EXPLANT STERILIZATION

Suitable sized plant material is sterilized as follows:

i. Clean the working area with ethanol, and start the air flow of the laminar air flow bench.
ii. Sterilized petridishes, distilled water, scalpel, filter paper sheets (suitable size and autoclaved), alcohol, disinfectant (mercuric chloride 0.01–0.1% aqueous solution, or 20% sodium hydrochloride), forceps, bead sterilizer, beaker or

coupling jar, spirit lamp or gas burner are collected on Laminar flow bench. Switch on the UV light of the bench for 30 minutes.

iii. Bring the plant material to be sterilized on the bench and prepare pieces of plant material for sterilization.

iv. Clean the working area and hands with alcohol, put on mask and cap, and lit the spirit lamp.

v. Keep 3–4 Petridishes in a row, add disinfectant in 1st plate and autoclaved distilled water in subsequent plates.

vi. Place plant pieces in 1st plate and immerse the material with the help of forceps for 5–10 minutes depending upon the disinfectant used.

vii. Transfer material from 1st to 2nd plate, rinse gently and pass to 3rd and 4th plate, one by one with thorough rinsing.

viii. Finally, drain the distilled water or place the material in a fresh Petridish or filter paper and prepare suitable sized explant.

5.4 CALLUS CULTURE

Cell from any plant species can be cultured aseptically on a nutrient medium. The cultures are initiated by planting an explant on an agar medium. Within 2–4 weeks, depending upon plant species, a mass of unorganized cells (callus) is produced. This callus can be subcultured indefinitely by transferring a small piece on to the fresh agar medium for further growth.

Juvenile tissues are generally most likely to produce a callus. However, callus cultures have been obtained from seedlings, young shoots or buds, root tips, developing embryos, fruits, floral parts, tubers and bulbs. Any plant tissue with living cells can be used as an explant after proper sterilization as described earlier. Under the influence of plant growth regulators, the cells of the explant may be induced to divide and form a loose mass of cells called as 'callus'. Callus tissue has also been obtained from wood phloem, leaf materials, fruits and other tissues.

When a suitable sized and sterilized explant is transferred aseptically under a laminar air flow bench on to a suitable nutrient medium (for example, MS medium) containing appropriate combination of plant growth regulators, it produces callus (Fig. 5.2). The callus is produced from outer layers of cortical cells in a stem explant by repetitive division of the cells. These dividing cells generate a pressure on the epidermis, which ultimately ruptures exposing newly formed callus. The continuous division of cells produces a mass of cells or callus on the explant. This callus is separated from the explants and transferred on to fresh medium. After attaining growth, callus is subcultured at regular interval of 3 to 4 weeks by dividing it into pieces (250 to 500 mg each) and transferred on to fresh medium. The callus piece used to inoculate a flask is known as 'inoculum'. Inocula (plural of inoculum) of similar shapes and sizes are used during the subcultures.

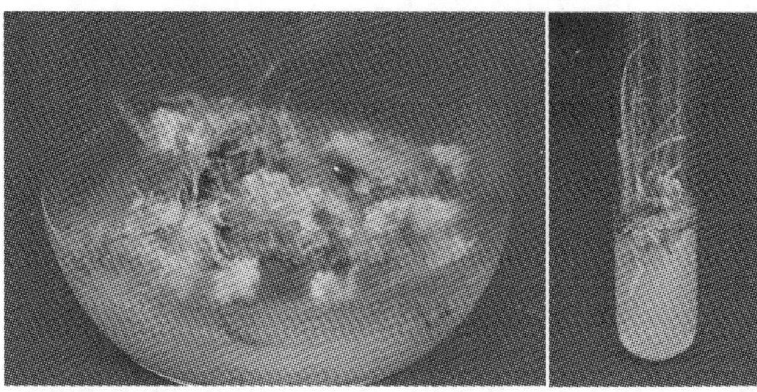

Fig. 5.2: Formation of callus from leaf explants (left) and regeneration of shoot cultures from callus (right).

Callus cultures are slow growing as compared to cell suspension cultures. Cells grow as clumps or masses in callus cultures and only lower cells are in contact with the medium whereas, cells in upper layers get their nutrients from cells in lower layers. Cells are in close association as compared to cell suspension cultures in which all the cells are in direct contact with the medium.

Applications of Callus Cultures

Callus cultures are slow growing systems on static medium and offer a unique system (as compared to *in vivo* grown plants) to study the following aspects of plant metabolism and differentiation:

1. Nutrition of plants.
2. Cell and organ differentiation and morphogenesis.
3. Somaclonal variation and its exploitation.
4. Developing cell suspension cultures and protoplasts cultures.
5. Genetic transformation using ballistic particle gun technology.
6. In the production of secondary metabolites and their regulation.

5.5 CELL SUSPENSION CULTURE

Cell suspension cultures are initiated by transferring friable callus to liquid nutrient medium. In liquid medium, plant tissues remains submerged which leads to anaerobic conditions and ultimately death of the cells. Therefore, such cultures are agitated on rotary shaker at 80–150 rpm, with an orbital diameter of 2.5–5.0 cm. Agitation serves both to aerate the cultures and to disperse the cells. Cells from the inoculum are separated during this process and a suspension of cells is obtained. The division rates of suspension culture cells at the exponential phase are typically higher than callus cells, but doubling times are slower than bacterial cells and usually vary from 24–72 hrs. It is a common observation that if relatively small number of cells are transferred (low inoculum density) to a new medium (either static or liquid), they may fail to divide; whereas a larger quantity of tissue transferred from the same culture may proliferate rapidly on the same medium. This observation has led to the concept of 'critical initial cell density'. This is defined at the smallest inoculum per volume of medium, from which a new culture can be reproducibly grown.

Following conditions determine the critical initial density of cells.

1. The cultures physiological characteristics.
2. The length of time and conditions under which the culture was previously maintained.
3. The composition of the fresh medium.

The third point is of interest. As the isolated cells failed to grow on fresh medium, 'conditioned medium' or 'nurse tissue' conditions are used to grow isolated cells or protoplasts. A 'conditioned medium' is the medium on which callus tissues were previously grown. Conditioning makes the minor adjustment in the nutrients and chemical substances released in the medium by the callus and promotes the growth of isolated cells of protoplasts. In cell suspension cultures, cells grow as isolated single cell and cell aggregates of a few cells to a few hundred cells. Cells aggregation varies from species to species and sometimes it is difficult to maintain fine cell suspension culture, *e.g.*, in members of the family Rutaceae.

Cell cultures are subcultured by dilution of the stock cultures, 5 to 10 times (v/v), depending upon the growth of cells. As mentioned above, growth of cell suspension cultures is always higher than callus cultures and, therefore, requires rapid subculture (7–21 days) as compared to callus cultures (4–8 weeks). Cell cultures in the liquid medium are grown as:

A. Batch cultures in shake flasks and bioreactor.
B. Continuous culture system.

5.5.1 Batch Culture

Whether cultures are grown in flasks (100 ml to 2 litres capacity containing 20 to 500 ml liquid medium) or in a bioreactor, system is used as a batch. Inoculated cultures are harvested after a definitive growth period (determined by growing cultures earlier and recording the growth phases during 1 to 6 weeks growth) at the end of stationary phase and cultures are harvested for a specific objective. Such cultures are grown again and again in batches for the purpose of experiments and known as batch cultures (Fig. 5.3).

Continuous cultures are usually grown in a bioreactor (Fig. 5.4). When inputs (nutrients, oxygen etc) and outputs (spent medium, cells) are precisely controlled and cells are grown under defined condition of nutrients, pH oxygen and cell density continuously for the purpose of study, such cultures are known as continuous cultures. This type of work is possible only after basic study of growth and metabolism in small shake flasks. In continuous culture, cells are always in an exponential phase of growth.

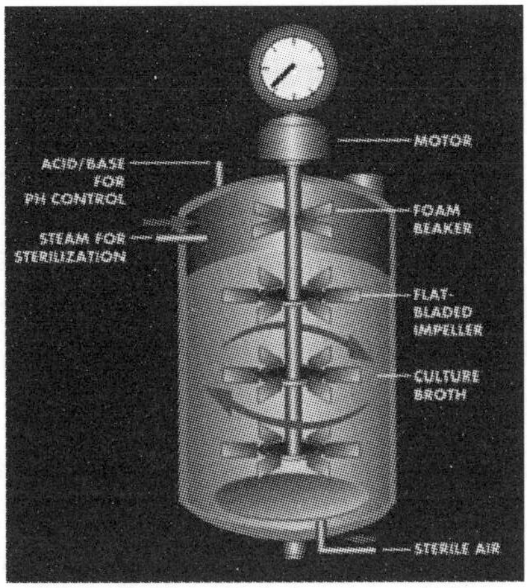

Fig. 5.3: Batch culture

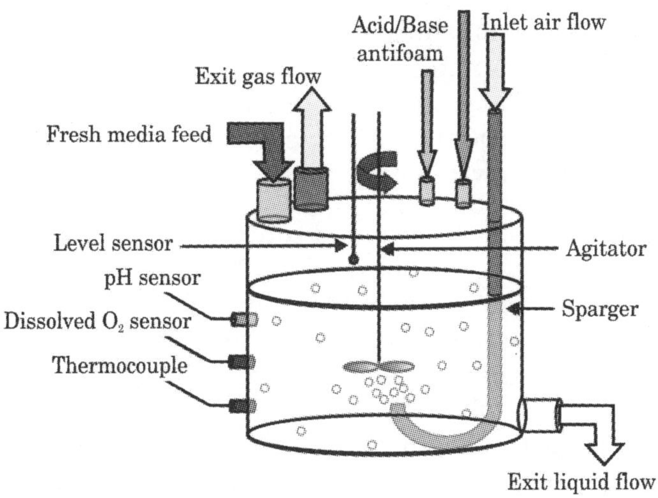

Fig. 5.4: Continuous culture

5.6 TIME COURSE OF GROWTH

The growth of undifferentiated callus and suspension culture can be quantified in a number of ways, but most commonly it is done as an increase in fresh weight, dry weight or cell numbers over a time course. Growth pattern of such cultures is always forming a sigmoid curve. This curve consists of a lag phase, in which there is a little or no cell division, an exponential phase in which maximum cell growth occurs, followed by a linear and stationary phase, in which the proportion of dividing cells gradually declines. Each stage can be characterized by distinct structural and biochemical features but, in fact, whole growth cycle is a continuous physiological process. At any particular time a heterogeneous population of cells differing in morphological and physiological characters can be observed. Usually the cultures are subcultured at the end of exponential phase or during stationary phase.

Fresh weight of callus is determined by carefully removing the adhering agar from callus and immediately weighing the callus in a sensitive balance. Cell cultures are filtered in a butchner funnel lined with filter paper, washed with distilled water (under mild vacuum) and weighed. Dry weights are determined by drying the cells to a constant weight at 60°C in an oven and weighing in a balance. Packed cell volume (PCV) of cell culture is determined by centrifuging it at 1000 rpm for 5 min in a graduated centrifuge tube and measuring the volume occupied by the cells out of total volume of the cell suspension. Cells are counted by a haemocytometer under compound microscope during growth period.

Different fractions so obtained contain single cells and clumps of cells depending upon the mesh size. These fractions can be used directly for initiating cell suspensions or may be used after concentrating the fraction by centrifugation. Fractions may be used to raise clones by plating the cells of these fractions. The appropriate fractions contain cells of similar shape, size and metabolic state. Such cultures are used to obtain synchronous cultures. Synchronous cultures are those cultures in which all the cells divide at the same time. Therefore, at one time all the cells divide and next time all the cells are in a stationary phase. This time, interval between two cycles of cell division is called 'doubling time'. Such cultures are useful in developing embryonic cultures as embryos of similar age can be obtained, which is also a necessity for the production of seed at large–scale of a species.

Cell plating and cell sieving are alternative methods to obtain strains and clones of cells. Microscopic observations and precautions are necessary to ensure that a colony of cells is originating from a single cell. By this method a true single cell clone can be obtained. Cell plating and cell sieving methods have been extensively used to obtain a large number of single cells and their colonies. Bergman (1959–1960) was the pioneer in establishing the cell plating technique.

5.7 MICROCULTURE CHAMBER

Another method of isolating and establishing single cell clones of higher plant cells is the microculture chamber method. This technique was developed in the laboratory of Prof. A.C. Hildebrandt at Wisconsin University, Madison, USA by Jones *et al.* (1960). For the study of single isolated cells with this method, the single cell is cultured aseptically in a drop of liquid medium surrounded by sterile mineral oil on a microscope slide. A mini chamber is prepared using three cover glasses as shown in the Fig. 5.5. The drop of medium containing cell/s is placed on a sterile microscope slide and ringed with sterile mineral oi1. All the cover glasses are sealed with mineral (Paraffin) oil ensuring aseptic growth. This micro–chamber is placed in a sterile Petri dish and incubated. Cells grow in this system and

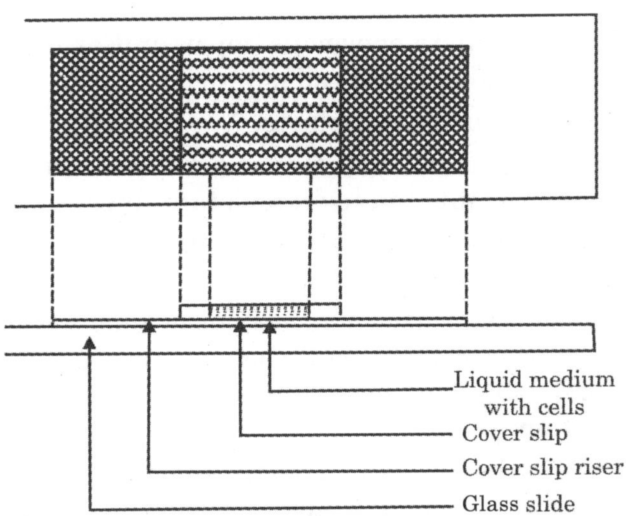

Fig. 5.5: Microculture chamber

can be observed periodically. Mineral oil prevents the water (medium) loss. Small colonies thus developed in chamber are aseptically transferred to fresh medium.

5.8 MEDIUM FOR CULTURE

Any plant tissue culture medium as described in Chapter 4 can be used to initiate cultures from an explant. All the media for callus or cell suspension culture contain a combination of plant growth regulators. The appropriate concentration and combination for a species is determined empirically. Usually medium contains an auxin and cytokinin for development of callus. Cultures may be grown in light or darkness depending upon the objective and requirement. The first evidence of callus formation can be recognized by rupturing the epidermis of the explant. When sufficient callus is produced it is separated from the explant and grown separately on a fresh medium.

5.9 APPLICATIONS OF CELL CULTURE

- In cell cloning by the plating technique with or without specific treatment, *e.g.*, mutagens, amino acid analogues, etc.
- Cell culture offers enormous opportunities in the study of single cell and group of cells.
- In the isolation of protoplasts.
- Development of cell lines for various types of resistance, *e.g.*, salt and drought, and toxin resistant lines.
- In scale–up technology using bioreactors of various types.
- In providing an excellent system of micropropagation on mass–scale by somatic embryogenesis in cell culture and use of bioreactor. Such system can be used for artificial seed production.
- In the process of cell division and factors affecting cell divisions and related process.
- In the study of nutrition.

- In cellular differentiation, single cell constitute an excellent system to study cytodifferentiation in a cell leading to tracheiry element formation and plastids differentiation etc.
- In secondary product formation, regulation and biosynthesis.

Chapter 6

Types of Plant Tissue Culture

6.1 PLASTICITY AND TOTIPOTENCY

Two concepts, plasticity and totipotency, are central to understanding plant cell culture and regeneration. Plants, due to their sessile nature and long life span, have developed a greater ability to endure extreme conditions and prediction than have animals. Many of the processes involved in plant growth and development adapt to environmental conditions. This plasticity allows plants to alter their metabolism, growth and development to best suit their environment. Particularly important aspects of this adaptation, as far as plant tissue culture and regeneration are concerned, are the abilities to initiate cell division from almost any tissue of the plant and to regenerate lost organs or undergo different developmental pathways in response to particular stimuli. When plant cells and tissues are cultured *in vitro* they generally exhibit a very high degree of plasticity, which allows one type of tissue or organ to be initiated from another type. In this way, whole plants can subsequently be regenerated (Fig. 6.1).

This regeneration of whole organisms depends upon the concept that all plant cells can, given the correct stimuli, express the total genetic potential of the parent plant. This maintenance of genetic potential is called 'totipotency'. Plant cell culture and regeneration do, in fact, provide the most compelling evidence for totipotency. In practical terms though, identifying the culture conditions and stimuli

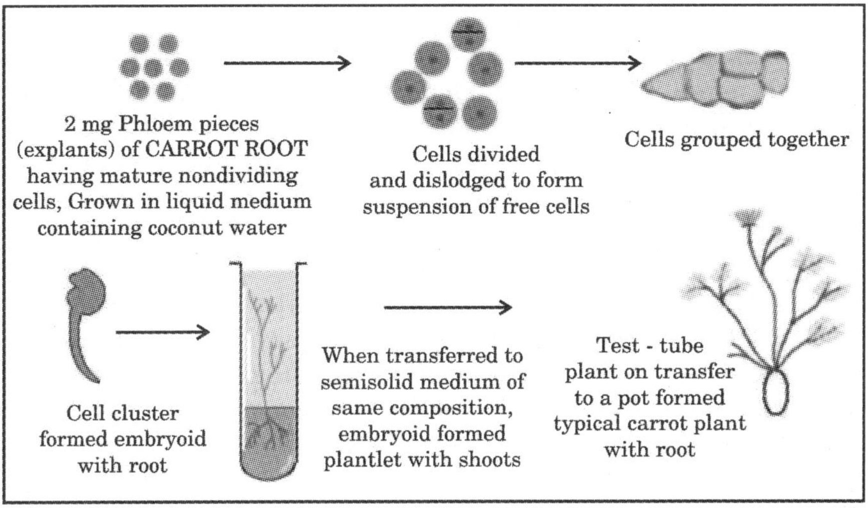

Fig. 6.1: Steward and co–worker's experiment showing totipotency of cells in carrot

required to manifest this totipotency can be extremely difficult and it is still a largely empirical process.

6.2 CULTURE TYPES

Cultures are generally initiated from sterile pieces of a plant. These pieces are termed 'explants', and may consist of pieces of organs, such as leaves or roots, or may be specific cell types, such as pollen or endosperm. Many features of the explant are known to affect the efficiency of culture initiation. Generally, younger, more rapidly growing tissue (or tissue at an early stage of development) is most effective.

Several different culture types most commonly used in plant regeneration studies will now be examined in more detail as follows:

6.2.1 Callus Culture

Explants, when cultured on the appropriate medium usually with both an auxin and a cytokinin, can give rise to an unorganized, growing and dividing mass of cells which is called as callus. It is thought that any plant tissue can be used as an explant, if the correct conditions are given. In culture, this proliferation can be maintained more or less indefinitely, provided that the callus is subcultured on a fresh medium periodically. During callus formation there is some degree of dedifferentiation (*i.e.*, the changes that occur during development and specialization are, to some extent, reversed), both in morphology (callus is usually composed of unspecialized parenchyma cells) and metabolism. One major consequence of this differentiation is that most plant cultures lose the ability to photosynthesis. This has important consequences for the culture of callus tissue, as the metabolic profile will probably not match that of the donor plant. This necessitates the addition of other components–such as vitamins and, most importantly, a carbon source–to the culture medium, in addition to the usual mineral nutrients.

Callus culture is often performed in the dark (the lack of photosynthetic capability being no drawback) as light can encourage differentiation of the callus. During long–term culture, the culture may lose the requirement for auxin and/or cytokinin. This process, known as 'habituation', is common in callus cultures from some plant species (such as sugar beet).

Callus cultures are extremely important in plant biotechnology. Manipulation of the auxin to cytokinin ratio in the medium can lead to the development of shoots, roots or somatic embryos from which whole plants can subsequently be produced. Callus cultures can also be used to initiate cell suspensions, which are used in a variety of ways in plant transformation studies. Various steps involved in plant regeneration from leaf callus culture in sugarcane have been demonstrated in Fig. 6.2.

A B C D

Fig. 6.2: Plant regeneration through leaf sheath callus

6.2.2 Cell Suspension Cultures

Callus cultures, broadly speaking, fall into one of two categories: compact or friable. In compact callus the cells are densely aggregated, whereas in friable callus the cells are only loosely associated with each other and the callus becomes soft and breaks apart easily. Friable callus provides the inoculum to form cell–suspension cultures. Explants from some plant species or particular cell types tend not to form friable callus, making cell–suspension initiation a difficult task. The friability of callus can sometimes be improved by manipulating the medium components or by repeated subculturing. The friability of the callus can also sometimes be improved by culturing it on 'semi–solid' medium (medium with a low concentration of gelling agent).

When friable callus is placed into a liquid medium (usually the same composition as the solid medium used for the callus culture) and then agitated, single cell and/or small clumps of cells are released into the medium. Under the correct conditions these released cells continue to grow and divide, eventually producing a cell–suspension culture. A relatively large inoculum should be used when initiating cell suspensions so that the released cell numbers build up quickly. The inoculum should not be too large though, as toxic products released from damaged or stressed cells can build up to lethal levels. Large cell clumps can be removed during subculture of the cell suspension.

Cell suspensions can be maintained relatively simply as batch cultures in conical flasks. They are continuously cultured by repeated subculturing into fresh medium. This results in dilution of the suspension and the initiation of another batch growth cycle. The degree of dilution during subculture should be determined empirically for each culture. A high degree of dilution will result in an extended lag period or, in extreme cases, death of the transferred cells.

After subculture, the cells divide and the biomass of the culture increases in a characteristic fashion, until nutrients in the medium are exhausted and/or toxic by–products build up to inhibitory levels–this is called the 'stationary phase'. If cells are

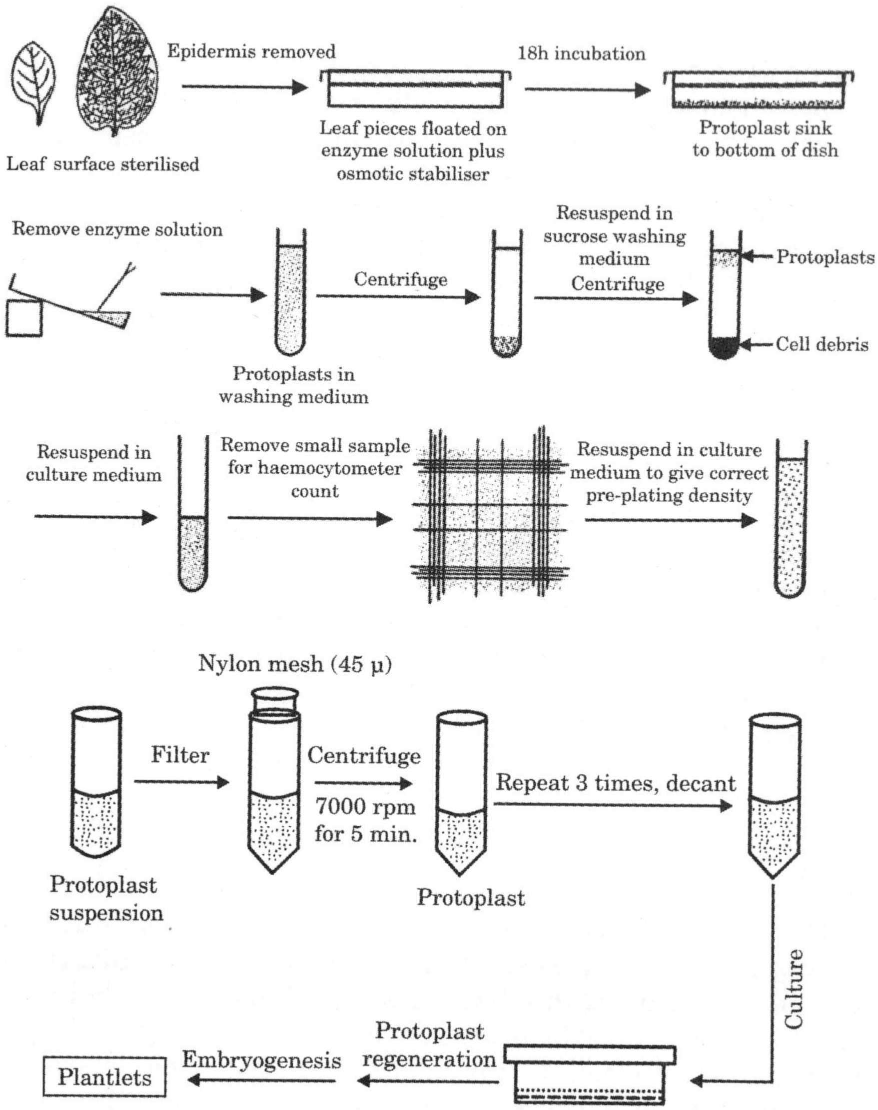

Fig. 6.3: Protoplast culture

left in the stationary phase for too long, they will die and the culture will be lost. Therefore, cells should be transferred as they enter the stationary phase. It is therefore, important that the batch growth–cycle parameters are determined for each cell–suspension culture.

6.2.3 Protoplast Culture

Protoplasts are plant cells with the cell wall removed. Protoplasts are most commonly isolated from either leaf mesophyll cells or cell suspensions, although other sources can be used to advantage. Two general approaches to remove the cell wall (a difficult task without damaging the protoplast) can be taken by mechanical or enzymatic degradation of cell wall.

Mechanical isolation, although possible, often results in low yields, poor quality and poor performance in culture due to substances released from damaged cells. Enzymatic isolation is usually carried out in a simple salt solution with a high osmoticum, plus the cell wall degrading enzymes. It is usual to use a mix of both cellulose and pectinase enzymes, which must be of high quality and purity.

Protoplasts are fragile and easily damaged, and therefore, must be cultured carefully. Liquid medium is not agitated and a high osmotic potential is maintained, at least in the initial stages. The liquid medium must be shallow enough to allow aeration in the absence of agitation. Protoplasts can be plated out on to solid medium and callus produced. Whole plants can be regenerated by organogenesis or somatic embryogenesis from this callus. Protoplasts are ideal targets for transformation by a variety of means.

6.2.4 Root Cultures

Root cultures (Fig. 6.4) can be established *in vitro* from explants of the root tip of either primary or lateral roots and can be cultured on fairly simple media. The growth of roots *in vitro* is potentially unlimited, as roots are indeterminate organs. Although the establishment of root cultures was one of the first achievements of modern plant tissue culture, they are not widely used in plant transformation studies.

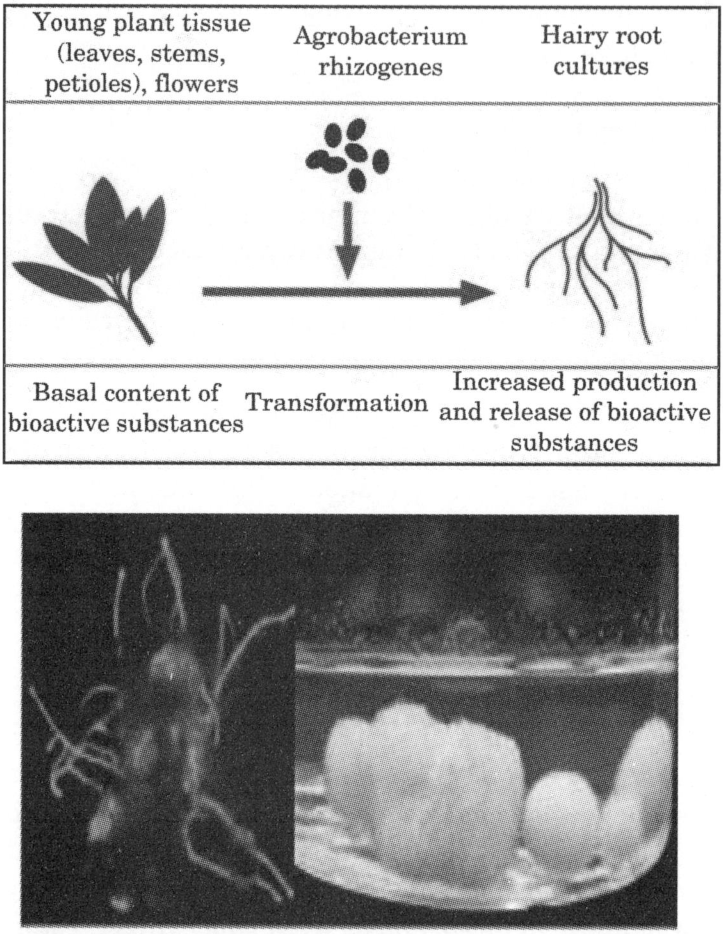

Fig. 6.4: Root culture

6.2.5 Shoot Tip and Meristem Culture

The tips of shoots (which contain the shoot apical meristem) can be cultured *in vitro,* producing clumps of shoots from axillary buds. This, method can be used for clonal propagation.

Shoot meristem cultures are potential alternatives to the more commonly used methods for cereal regeneration as they are less genotype–dependent and more efficient.

6.2.6 Embryo Culture

Embryos can be used as explants to generate callus cultures or somatic embryos. Both immature and mature embryos can be used as explants. Immature, embryo–derived embryogenic callus is the most popular method of monocot plant regeneration.

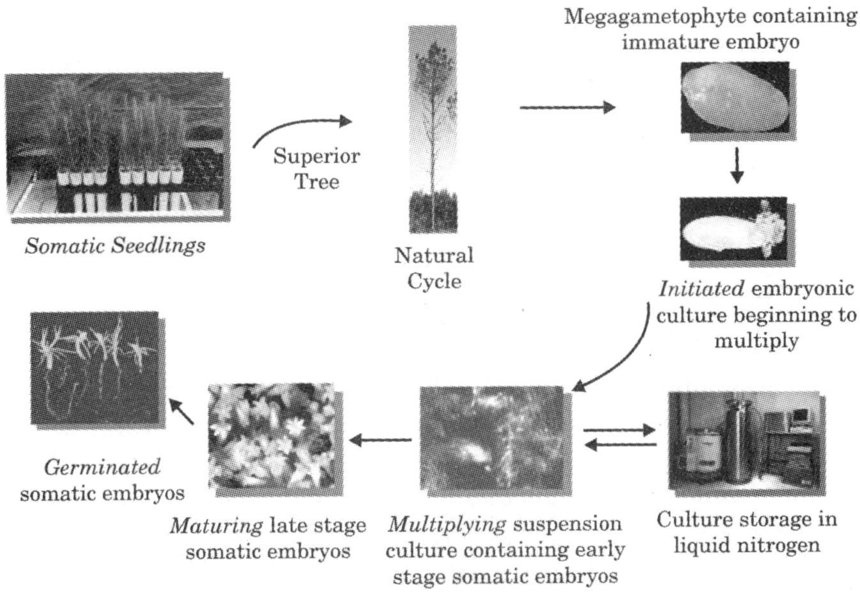

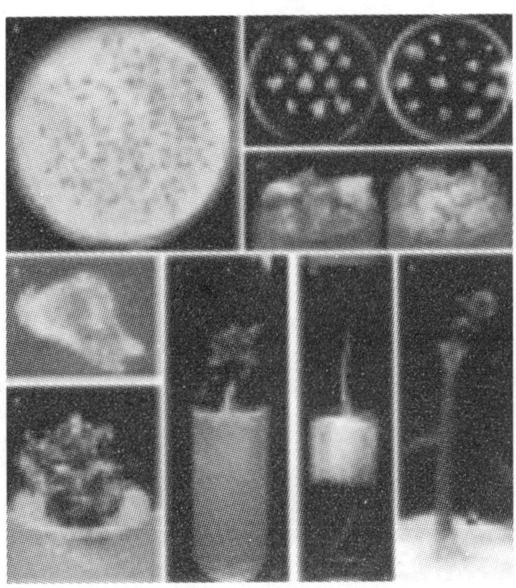

Fig. 6.5: Embryo culture

6.2.7 Microspore Culture

Haploid tissue can be cultured *in vitro* by using pollen or anthers as an explant. Pollen contains the male gametophyte, which is termed the 'microspore'. Both callus and embryos can be produced from pollen. Two main approaches can be taken to produce *in vitro* cultures from haploid tissue.

The first method depends on using the anther as the explant. Anthers (somatic tissue that surrounds and contains the pollen) can be cultured on solid medium (agar should not be used to solidify the medium as it contains inhibitory substances). Pollen–derived embryos are subsequently produced via dehiscence of the mature anthers. The dehiscence of the anther depends both on its isolation at the correct stage and under the correct culture conditions. In some species, the reliance on natural dehiscence can be circumvented by cutting the wall of the anther, although this does, of course, take a considerable amount of time. Anthers can also be cultured in liquid medium, and pollen released from the anthers can be induced to form embryos, although the efficiency of plant regeneration is often very low. Immature pollen can also be extracted from developing anthers and cultured directly, although this is a very time–consuming process (Fig. 6.6).

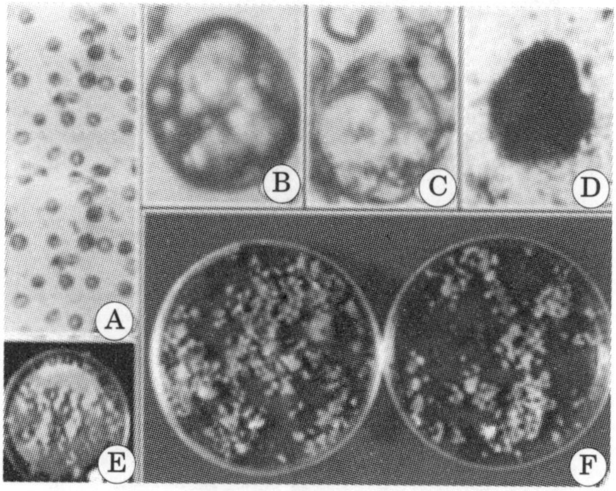

Fig. 6.6: Microspore culture

Both methods have advantages and disadvantages. Some beneficial effects to the culture are observed when anthers are used as the explant material. There is, however, the danger that some of the embryos produced from anther culture will originate from the somatic anther tissue rather than the haploid microspore cells.

If isolated pollen is used there is no chance of mixed embryo formation, but the efficiency is low and the process is time–consuming.

In microspore culture, the condition of the donor plant is of critical importance, as is the timing of isolation. Pretreatments, such as a cold treatment, are often found to increase the efficiency. These pretreatments can be applied before culture, or in some species, after placing the anthers in culture.

Plant species can be divided into two groups, depending on whether they require the addition of plant growth regulators to the medium for pollen/anther culture; though they also often require organic supplements, *e.g.*, amino acids. Any of the cereals (rice, wheat, barley and maize) require medium supplemented with plant growth regulators for pollen/anther culture. Regeneration from microspore explants can be obtained by direct embryogenesis, or *via.*, a callus stage and subsequent embryogenesis. Haploid tissue cultures can also be initiated from the female gametophyte (the ovule). In some cases, this is a more efficient method than using pollen or anthers.

The plants obtained from haploid cultures may not be always haploid. This can be a consequence of chromosome doubling during the culture period. Chromosome doubling (which often has to be induced by treatment with chemicals such as colchicine) may be an advantage, as in many cases haploid plants are not the desired outcome of regeneration from haploid tissues. Such plants are often, referred to as 'di–haploids', because they contain two copies of the same haploid genome.

6.3 PLANT REGENERATION

In broad terms, two methods of plant regeneration are widely used in plant regeneration studies, *i.e.*, somatic embryogenesis and organogenesis.

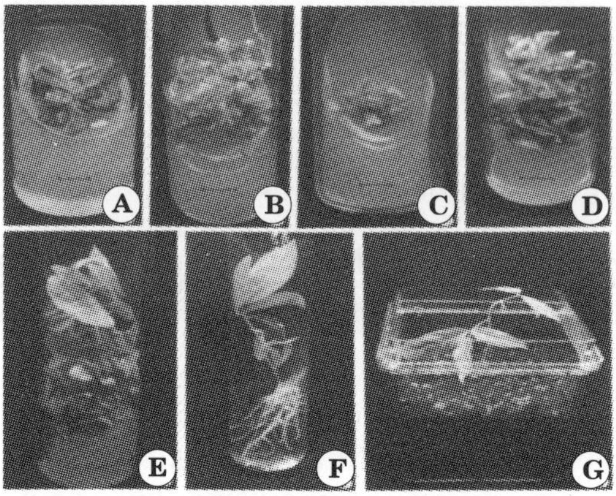

Fig. 6.7: Plant regeneration

6.3.1 Somatic Embryogenesis

In somatic (asexual) embryogenesis, embryo–like structures, which can develop into whole plants in a way analogous to zygotic embryos, are formed from somatic tissues (Fig. 6.8). These somatic embryos can be produced either directly or indirectly. In direct somatic embryogenesis, the embryo is formed directly from a cell or small group of cells without the production of an intervening callus. Though common from some tissues (usually reproductive tissues such as the nucellus, styles or pollen), direct somatic embryogenesis is generally rare in comparison with indirect somatic embryogenesis.

In indirect somatic embryogenesis, callus is first produced from the explants, embryos can then be produced from the callus tissue or from a cell suspension produced from that callus. Somatic embryogenesis from carrot is the classical example of indirect somatic embryogenesis.

Somatic embryogenesis usually proceeds in two distinct stages. In the initial stage (embryo initiation), a high concentration of 2, 4–D is used. In the second stage (embryo production) embryos are produced in a medium with no or very low levels of 2, 4–D.

6.3.1.1 *Indirect somatic embryogenesis in carrot (Daucus carota)*

A callus can be established from explants from a wide range of carrot tissues by placing the explant on solid medium (*e.g.*, Murashige and

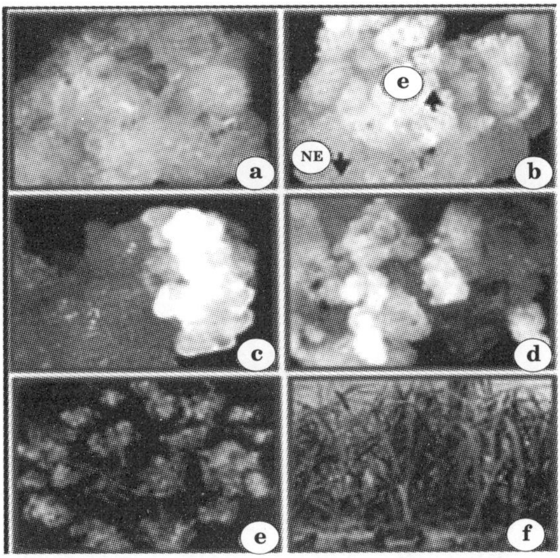

Fig. 6.8: A schematic representation of the sequential stages of somatic embryo development. Somatic embryos may develop from single cells or from a small group of cells. Repeated cell divisions lead to the production of a group of cells that develop into an organised structure known as a 'globular–stage embryo'. Further development results in heart and torpedo shaped embryos, from which plants can be regenerated. Zygotic embryos undergo a fundamentally similar development through the globular (which is formed after the 16–cell stage), heart and torpedo stages. Polarity is established early in embryo development. Signs of tissue differentiation become apparent at the globular stage and apical meristems are apparent in heart–stage embryos.

Skoog (MS) containing, 2, 4–D (1 mg 1^{-1}). This callus can be used to produce a cell suspension by placing it in agitated liquid MS medium containing 2, 4–D (1 mg 1^{-1}). This cell suspension can be maintained by repeated subculturing into 2, 4–D containing medium. Removal of the old 2, 4–D containing medium and replacement with fresh medium containing abscisic acid (0.025 mg 1^{-1}) results in the production of embryos.

6.3.1.2 *Direct somatic embryogenesis in alfalfa (Medicago falcata)*

Young trifoliate leaves are used as the explant. These are removed from the plant and chopped into small pieces. The pieces are washed in a

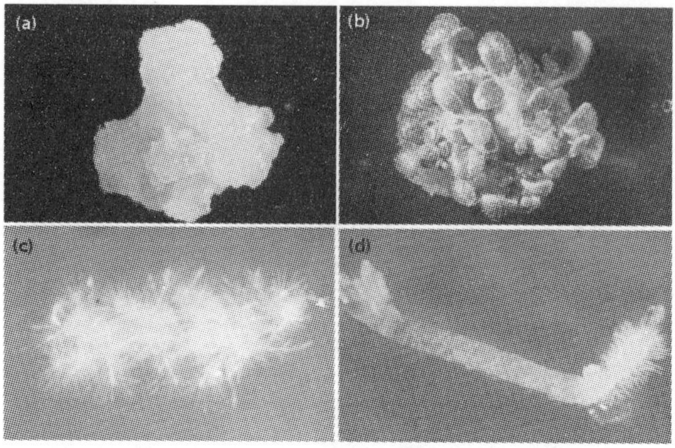

Fig. 6.9: Growth in plant tissue cultures can take many forms ranging from (a) disorganised callus, in this case initiated from carrot hypocotyl on a nutrient medium supplemented with 0.45 µM 2,4–D. (b) Shoot regeneration from a callus derived from a single protoplast of *Hyoscyamus muticus*. (c) Root regeneration from a callus derived from a root explant of *Arabidopsis thaliana*. (d) A single somatic embryo removed from a callus derived from a carrot cotyledon explant.

plant growth regulator–free medium and placed in liquid medium B_5 supplemented with 2, 4–D(4 mg 1^{-1}), kinetin (0.2 mg 1^{-1}), adenine (1 mg 1^{-1}) and gluathione (10 mg 1^{-1}). The cultures are maintained in agitated liquid medium for about 10–15 days. Washing the explants and replacing the old medium with B_5 medium supplemented with maltose and polyethylene glycol results in the development of the somatic embryos. These somatic embryos can be matured on solid medium containing abscisic acid. Note that in both cases, although the production of somatic embryos from alfalfa necessitates the use of more complicated media, the production of embryos is fundamentally a two–step process. The initial medium, which contains 2, 4–D, is replaced with a medium that does not contain 2, 4–D. In many systems it has been found that somatic embryogenesis is improved by supplying a source of reduced nitrogen, such as specific amino acids or casein hydrolysate.

6.3.2 Case Study on Regeneration *via.*, Somatic Embryogenesis from Immature or Mature Embryos in Cereal Species

The principal method adopted for the tissue culture and regeneration of a wide range of cereal species is somatic embryogenesis, using cultures initiated from immature zygotic embryos. Embryogenic callus is normally initiated by placing the immature embryo on to a medium containing 2, 4–D. Shoot regenerations is initiated by placing the embryogenic callus on a medium with BAP (with or without 2, 4–D). These shoots can be subsequently rooted. The medium used for induction of embryogenic callus is usually a modified MS (for *Triticale*) or n6 (for rice and maize). Maltose is used as the other carbon source in preference to sucrose and additional organic supplements (such as specific amino acids, yeast extract and/or casein hydrolysate) are common. The isolation and culture of immature embryos is, however a labour–intensive and relatively expensive procedure. An additional problem is the small target size if the immature embryos are to be used for biolistic transformation. Alternatives are therefore being sought. One alternative is to use mature embryos (or seeds) as the explant to initiate embryogenic callus. This approach has been successfully applied to several cereal species such as rice and oats. The culture techniques and media used for culture establishment and regeneration from mature embryo/seed–derived cultures are fundamentally the same as those used for cultures initiated from immature embryos.

6.4 THE IMPORTANCE OF GENOTYPE

The major influence on tissue–culture response appears to be genetic, with culture requirements varying between species and cultivars. Model genotypes that responded well to culture *in vitro* were initially used in plant transformation studies. However, most of the model genotypes used were not elite commercial cultivars. The commercial cultivars tended to respond poorly to culture *in vitro*. One of the main aims is therefore to identify the components that make up a widely applicable, optimal culture regime. Many factors have been investigated for their ability to improve the culture response from elite cultivars, including media components (such as alternative carbon sources, macro–and micro elements concentrations and composition), media preparation method, donor plant condition and growth conditions.

6.5 ORGANOGENESIS

Somatic embryogenesis relies on plant regeneration through a process analogous to zygotic embryo germination. Organogenesis relies on the production of organs, either directly from an explant (Fig. 6.10) or from a callus culture. There are three methods of plant regeneration *via.*, organogenesis. The first two methods depend on adventitious organs arising either from a callus culture or directly from an explant. Alternatively, axillary bud formation and growth can also be used to regenerate whole plants from some types of tissue culture.

Fig. 6.10: Organogenesis

Organogenesis relies on the inherent plasticity of plant tissues, and is regulated by altering the components of the medium. In particular, it is the auxin to cytokinin ratio of the medium that determines the developmental pathway of the regenerating tissue. It is usual to induce shoot formation by increasing the cytokinin to auxin ratio in the culture medium. These shoots can then be rooted relatively simply.

6.5.1 Organogenesis in Tobacco (*Nicotiana tabacum*)

Organogenesis from tobacco pith callus is the classical example of how varying plant growth regulator regimes can be used to manipulate the pattern of regeneration from plant tissue cultures.

When cultured on a medium containing both auxin and cytokinin, callus will proliferate. If the auxin to cytokinin ratio is increased, adventitious roots will form from the callus by organogenesis. If the auxin to cytokinin ratio is decreased adventitious shoots will be formed. If the explants are cultured on medium containing only a cytokinin shoots can be produced directly.

Tobacco plants can also be easily regenerated from tobacco leaf pieces. Leaves are cut into approximately 1 cm squares with a sterile scalpel (avoiding large leaf veins and any damaged areas). The leaf pieces are then transferred (right side up) to gelled MS medium supplemented with 1 mg 1^{-1} BAP (a cytokinin) and 0.1 mg 1^{-1} NAA (an auxin). Over the next few weeks, callus forms on the explants, particularly around the cut surfaces. After 3 to 5 weeks shoots emerge directly from the explants or from callus derived from the explants. When these shoots are about l cm long they can be cut at the base and placed on to solid MS medium without any plant growth regulator. The shoots will form roots and plantlets that will grow in this medium and can subsequently be transferred to soil.

6.6 APPLICATION OF TISSUE CULTURE INTO PLANTS

6.6.1 Transformation Protocols

Various methods of plant regeneration are available to the plant biotechnologist. Some plant species may be amenable to regeneration by a variety of methods, but some may only be regenerated by a specific method. In Chapter 3, various methods that can be used to transform plants have been considered, but it is worthwhile briefly considering the interaction of plant regeneration methodology and transformation methodology. Not all plant tissue is suited to every plant transformation method, and not all plant species can be regenerated by every method. It is therefore needed to consider both a suitable plant tissue culture/regeneration regime and a compatible plant transformation methodology for successful transformations.

Chapter 7

Sugarcane and Sugar

Sugarcane is one of the most efficient converters of solar energy into sugars and other renewable forms of energy. The plant was domesticated by the Polynesians (Brandes, 1958) for its sweet stem, but presently it has emerged as a multipurpose crop providing not only sugar but also a series of value added products such as paper, ethanol and other alcohol derived chemicals, animal feed, antibiotics, particle board, bio–fertilizer and raw material for generating electricity. Global sugar consumption has been increasing at a steady rate of 2 per cent per annum. Ethanol has emerged as a key product from the sugarcane industry globally. With ever increasing oil prices,

Fig. 7.1: Sugarcane field

more and more countries are encouraging plant–based ethanol production as an environment-friendly fuel (Fig. 7.1).

In a typical sugar mill, 100 tonnes (t) of sugarcane on an average produce 10 t of sugar, 4 t of molasses from which ethanol is produced, 3 t of press–mud which is converted into biofertilizer, 30 t of bagasse used for cogeneration of power to yield 1,500 kW electricity and for manufacturing paper. Apart from these, about thirty tonnes of cane tops and leaves are generally left in the field, which through recycling further add to the economic value of the crop. Sugarcane, thus, plays a major role in the economy of sugarcane growing areas and, hence, improving sugarcane production will greatly help in economic prosperity of the farmers and other stakeholders associated with sugarcane cultivation.

About 20 countries in the Asia–Pacific region grow sugarcane on a commercial basis contributing 608.37 million tonnes (mt) to the world production of 1,387.78 mt. However, sugarcane yields vary widely across the regions, ranging from 17.1 tonnes/hectare (t/ha) in Cambodia to 91.97 t/ha in Australia with an average yield of 56.66 t/ha compared to the world average of 67.98 t/ha. Most of the sugar–cane farmers in this region are small and confronted with problems of low cane yields due to poor quality seed, low fertilizer inputs, prevalence of diseases and pests, lack of proper irrigation facilities, untimely harvests and several other local constraints. The limited cultivable area available for expansion and continuing conversion of agricultural land for non–agricultural purposes necessitate that production increase comes mainly from increase in per hectare yields. Improved agronomic practices, use of required quantity of fertilizer at appropriate time, better irrigation facilities, comprehensive disease and pest management packages and regular development of improved varieties are the necessary inputs required for improving sugarcane production and productivity. Besides, availability of disease and pest–free, true to type planting material is an important prerequisite for achieving the desired yield improvement. Sugarcane, being a vegetatively propogated crop, has a low 1:6 to 1:8 seed multiplication rate. Hence, non–availability of seed material is one of the major problems faced by farmers in developing countries.

Further, the bulky cane cuttings used for planting as seed harbor many pests and diseases thereby decreasing cane yield and quality drastically. Accumulation of disease over vegetative cycles leads to further yield and quality over the years. In fact, poor quality seed is a major constraint in sugarcane production (Fig. 7.2).

Fig. 7.2: Sugarcane stems

Development of tissue culture technology for rapid multiplication of disease – free planting material has greatly facilitated mass production of quality seed in sugarcane. Micropropagation techniques have been adopted successfully by farmers and industry in some sugarcane growing countries of Asia–Pacific, *e.g.* India, Australia and the Philippines.

7.1 THE HISTORY OF SUGARCANE

Over 2000 years ago in the Pacific Islands, the stalks of a plant were found to contain a sweet liquid. The plant is now known as sugarcane. In the 17th century, another plant was found to contain sugar. It is called sugar beet. The number of operating sugar mills in the country has increased from 29 in 1930–31 to 412 by 1996–97 and a little above 475 in 2007–2008. Some 95 factories are, however, non–operational.

The extent of control over the industry has been very comprehensive and inhibitive. It is the Central Government that decides where and when a sugar factory can be opened. It decides what technology the sugar factory will deploy, what kind of packaging material it will use for bagging the sugar and at what prices the bags can be purchased, what wages it will pay to the factory workers as also to the labor employed for harvesting and transport. It decides, again, the percentage of total sugar production that the factories will deliver as levy sugar and at what price for the Public Distribution System, the Statutory Minimum Price to be paid to the cane growers, the free sale quota and the monthly releases thereof, what quantity could be imported/exported and at what price and at what tariffs. In fact, it may be said that there was hardly any decision that a management of the sugar factory, whether cooperative or private could take without some kind of intervention by the government.

In case of the cooperative mills the governmental control is even closer. The government decides on their credit and expansion plans as also diversification into byproducts. The cooperative factories have grown into stereotypes that follow the Open Pan technology and produce only standard varieties of sugar with little attention to byproducts. This is quite contrary to the general pattern in main producer countries of the world under which, sugar invariably was the byproduct and the derivatives the main revenue earners. The government rules were clearly influenced by political considerations and urban consumer interests. For example, it was ordained that the delivery of sugar cane by the members of the cooperative factory will be treated as a transaction of sale and purchase and, consequently, subject to a hefty purchase tax. The licenses for factories were doled out for political considerations. The Levy prices were fixed to favour the northern states.

7.2 SOME KEY FACTS

- Sugar is produced in more than 100 countries around the world;
- Over 25 nations produce sugar from sugar beets, around 80 from sugarcane, and another eight from both sugarcane and sugar beet;
- World sugar production for 2006–07 was 164 million tonnes;

- As many as 150 countries are engaged in sugar trade mostly as importers;
- World sugar trade accounts for 45–50 million tonnes;
- The closing stock of 41.5 million tonnes sugar for 2006–07 has been the highest in the recorded sugar history;
- Sugar is used as the primary feedstock in as much as 60 per cent of world ethanol production;
- In Brazil, half the total sugarcane output is diverted to ethanol production;
- India's per capita sugar consumption for 2005 stood at 19.6 kg.
- Swaziland's per capita sugar consumption of 97.4 kg in 2005 is the world's highest, closely followed by Switzerland and Israel at 70.3 kg and 65.7 kg respectively;
- Sugar consumption in Rwanda and Congo in 2005, at 1.5 kg and 1.6 kg per capita respectively, is the lowest in the world.
- A spoonful of sugar added to a vase prolongs the lives of freshly cut flowers;
- A pinch of sugar on the tongue is a traditional remedy for hiccups; and
- A teaspoon ful of sugar after a hot curry dulls the heat in your mouth.

7.3 INDIA'S SWEETENER

- The Indian Sugar industry accounted for around 1% of GDP of the country during the financial year 2009–10.
- The sugar industry has employed 0.5 million workers and also provides substantial indirect employment through various ancillary activities.
- For the year 2006–07, India produced 30.6 million tonnes of sugar; and in the year 2009–10 India produced 41.2 million tonnes of sugar;
- Its average sugar output over the last five years stood at 20.6 million tonnes; and
- The country's sugar consumption is estimated to be more than 20 million tonnes.

Chapter 8

Sugar and Human Health

India is supposed to be the home of sugarcane and sugar. Indians knew the art of making sugar since at least the fourth century. The advent of the modern sugar industry in India dates back only to mid 1930s when a few vacuum pan units were established in the sub–tropical states of Uttar Pradesh and Bihar. Until the mid–1950s, the sugar industry was almost wholly confined to these states. It was only after late 1950s/ early 60s that the industry dispersed into southern, western and other parts of northern India. India is the largest consumer and the second largest producer of sugar in the world (Fig. 8.1).

Fig. 8.1: Different types of sugar

Imagine the desserts like ice–creams, kulfi, jalebis, halwas, rasmalai and rosogollas before you, how hard it would be to resist the temptation. And later feel guilty counting calories. To set our basics right on these sugary delights here are some attributes we need to know about sugar and health. Sugars are a major form of carbohydrates and are found probably in all green plants. They occur in significant amounts in most fruits and vegetables. There are three main simple sugars *i.e.*, sucrose, fructose and glucose. Sucrose in fact is a combination of fructose and glucose and the body quickly breaks it down into simpler substances. Over 50 million farmers and their dependants as also a large number of agricultural labourers are involved in sugar cultivation, harvesting and ancillary activities.

A balanced diet can (and should) come from a variety of different foods, calculated to give the desired level of carbohydrates, proteins, fats, vitamins, and minerals. Nutritional scientists advocate that carbohydrates should provide at least 50% of energy requirements.

Apart from sugar being the cheapest instant source of energy, it has several medical and therapeutically values. Some of them are described as under:

- ***Sugar for oral rehydration*:** Sugar is extremely valuable in treatment of severe infantile diarrhea, a serious problem that kills around 3.5 million children in a year in under-developed countries.
- ***Sugar heals wounds*:** Sugar helps in wiping out infections in all types of wounds and speed up the healing as well. Sugar fills up open wounds and it dissolves in tissue where micro organisms cannot survive.
- ***Sugar is a good carrier of vitamins and minerals*:** Vitamin A deficiency in South American population is being combated successfully with the use of fortified sugar. Encouraged by the results, fortification of sugar with vitamin A is being attempted to combat vitamin A as well as mineral (iron) deficiencies a major area of concern in some of the developing countries.
- ***Empty nutrition myth*:** The complaint about sugar is that it contributes only calories but no nutrients to the diet. "Empty calorie" myth is really an empty complaint. As sugar is never eaten alone. It is almost always consumed as an essential or enhancing ingredient in a wide variety of other nutritious foods. Sugar is 100% utilizable carbohydrate.

8.1 SUGAR AND DIABETES

Diabetes "mellitus" is a disorder of the body's metabolism. The carbohydrates which a normal person eats are converted into glucose which is absorbed into the blood stream and is used by the body for energy (Fig. 8.2). The energy (calories) produced by different edibles is shown in Table 8.1.

Table 8.1: Calorie counter

Sugar (Serving size)	*Weight (g)*	*Calories*
Table Sugar, 1 level teaspoon	4	15
Besan Laddu	40	315
Carrot Halwa	135	460
Coconut Barfi	30	157
Kaju Rolls	14	83
Kheer Rice	150	345
Kulfi Malai	75	210
Kulfi Mango	70	147
Malpua	40	342
Naan Khatai	45	150
Rasgulla	40	194
Sandesh	7	30
Shrikhand	105	465
Til Laddu	20	55

Fig. 8.2: Different types of sweets

The diabetic person cannot utilize carbohydrates efficiently as a result blood has a higher than the normal amount of glucose and it is often present in urine. Though all the causes of diabetes are not

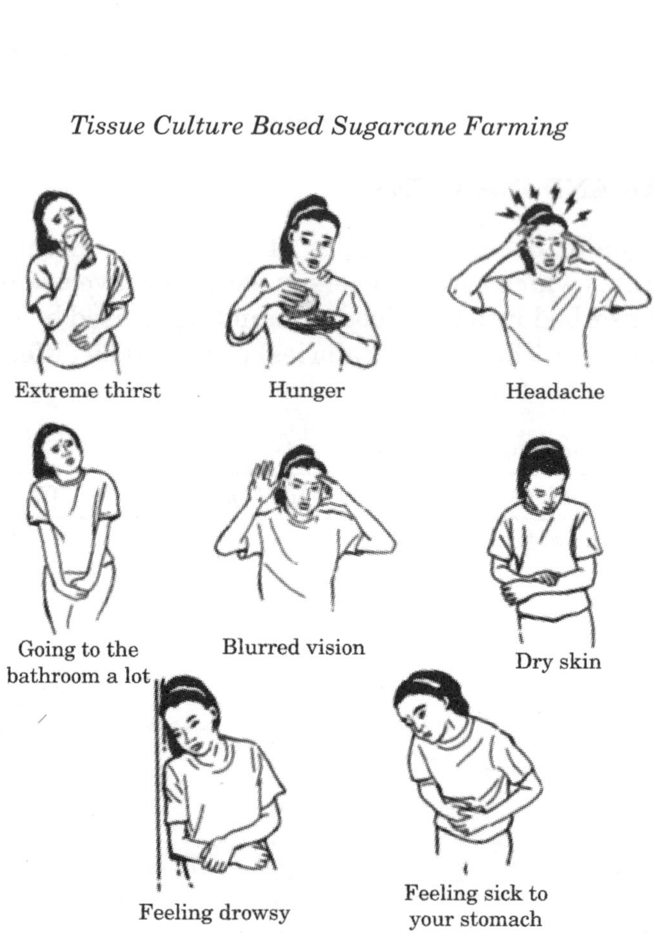
Extreme thirst
Hunger
Headache
Going to the bathroom a lot
Blurred vision
Dry skin
Feeling drowsy
Feeling sick to your stomach

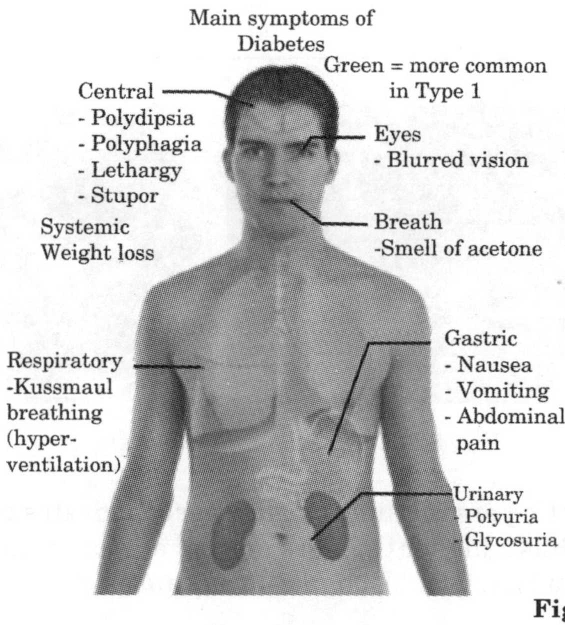
Main symptoms of Diabetes
Green = more common in Type 1
Central
- Polydipsia
- Polyphagia
- Lethargy
- Stupor
Eyes
- Blurred vision
Systemic
Weight loss
Breath
-Smell of acetone
Respiratory
-Kussmaul breathing (hyper-ventilation)
Gastric
- Nausea
- Vomiting
- Abdominal pain
Urinary
- Polyuria
- Glycosuria

Fig. 8.3: (*Contd...*)

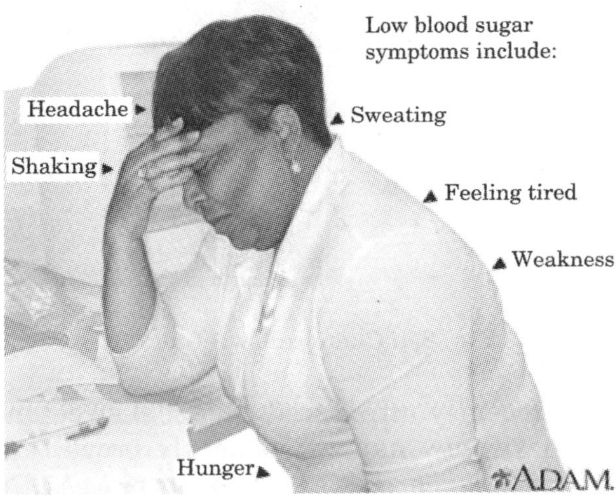

Fig. 8.3: Symptoms of low sugar in blood

known, genetic factors play a major part in the development of this disease. Once a person develops diabetes the intake of sugar and other simple carbohydrates is usually restricted. However, there is little actual evidence to support this practice. Research now indicates that not all starches and sugars are alike in their effect on blood sugar. For example carrots can produce a higher blood sugar than a candy bar and honey has a greater effect than sucrose. Eating modest amounts of sucrose as a part of regular meals does not lead to higher amount of blood sugar as has been commonly supposed. The dietary recommendations of the American dietary association (ADA) are in most respect the same as non–diabetic person and are based on sound principal of nutrition. Eating a modest amount of sugar is acceptable says (ADA) council of nutrition as long as metabolic control is maintained (Fig. 8.3).

8.2 CHEMICAL SWEETENERS

A number of chemical sweeteners have been produced and used from time to time. With passage of time and further research, some of these had to be banned as being harmful to health.

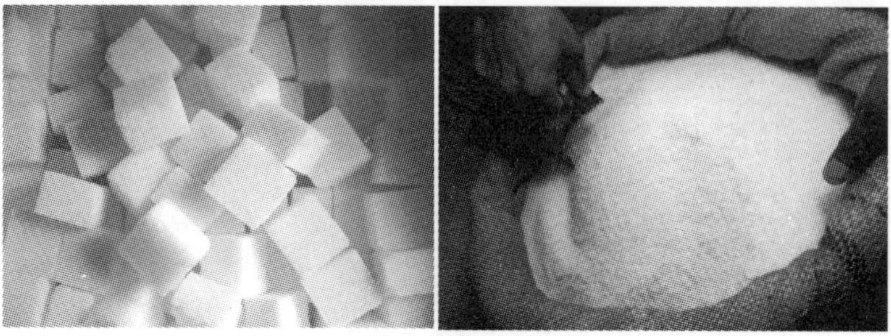

Fig. 8.4: Chemical sweetners

Some of the recently introduced chemical sweeteners such as cyclamatic, saccharin etc may subsequently prove to be a health hazard since scientific research is a continuous process. Consumption of such products may thus be beset with unknown risk which may become known later. On the other hand, sugar, whether expressed in terms of sucrose or fructose is a natural product and is a wonderful gift of nature to mankind. It has many virtues and cannot cause health hazard unless misused. Therefore, the risk evaluation between sugar consumption and that of chemical sweeteners must take this important factor into account (Fig. 8.4).

Chapter 9

Sugarcane Seed Production

Sugarcane stem, used as planting material (seed), is composed of a series of nodes and internodes (Fig. 9.1). Each node has a leaf, in the axil of which a bud is located. The bud has a dormant apical meristem well protected by several tightly clasping bud scales. Besides the bud, the node possesses a root band zone bordered by a growth ring. The root band contains one to several rows of root primordia which produces roots when the cane cuttings are planted. The growth ring is an intercalary meristem located immediately above the root band. Cane cuttings with one, two or three buds, known as "setts", "seed canes" or "seed pieces" are used as seed. In some instances, buds scooped out of the cane with a bud chipping machine are used for raising the seed nursery. For raising a healthy sugarcane crop, setts should be harvested from 7 to 10 months old crop which is totally free of diseases and pests. The setts should be healthy and must have high moisture content. The buds should be dormant and the

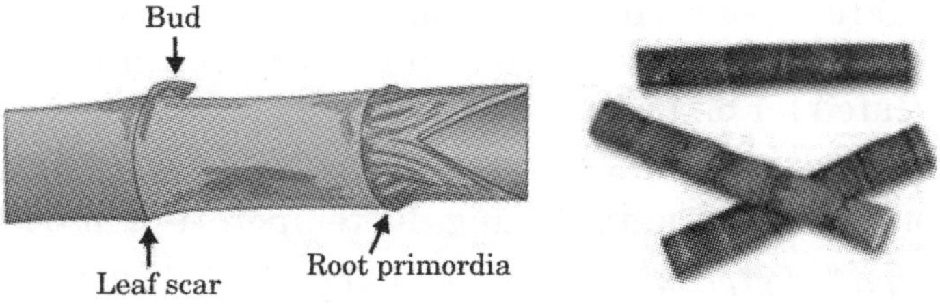

Fig. 9.1: Sugarcane stem

canes used to obtain seed setts must be free from rooting at the nodes, splits on the internodes and other damages.

9.1 SEED SETT PREPARATION

In India, seed setts are prepared manually. Seed canes are harvested and dry leaves removed manually to avoid any damage to the buds. Canes are cut with a sharp knife into setts containing two or three buds each. Sett–cutting machines have now become available making the process more efficient and convenient.

The cut ends of seed setts become easy entry points of many disease causing microbes, leading to sett rotting and damage to the buds and root primordia. Soaking the setts for 5 to 10 minutes in 0.1 per cent solution of a systemic fungicide such as methyl benzimidazole–2yl–carbamate (MBC) just before planting is recommended to ensure protection.

9.1.1 Heat Treatment of Setts

Sugarcane setts may harbour a host of diseases such as sugarcane smut, red rot, grassy shoot, ratoon stunting, sugarcane mosaic and yellow leaf. Also, scale insects and borers present on the setts can cause heavy damage to the new crop. Heat treatment of setts helps in getting rid of several diseases and pests.

There are four types of heat therapies:

(1) ***Hot water:*** Setts are immersed in water maintained at 50°C for two to two and a half hours. Often, fungicides are mixed in hot water to eliminate smut disease.

(2) ***Hot air:*** Dry heat produced by electric heaters placed at different points in the heating chamber is circulated with a fan. Temperature is maintained at 56°C and the seed is treated for eight hours.

(3) ***Moist hot air:*** Steam is injected into the treatment chamber for four hours maintaining the temperature at 54°C.

(4) ***Aerated steam:*** Steam is mixed with air in 1:4 proportions and forced into the treatment chamber through small holes. The treatment is given for one hour at 50°C.

When applied properly, heat therapy eliminates ratoon stunting disease, grassy shoot disease, sugarcane smut disease, and also seed borne insect pests.

9.2 SEED PRODUCTION

A three–tier seed production system comprising primary (breeder's) seed, secondary (foundation) seed and certified seed production as detailed below is ideally followed.

9.2.1 Primary (Breeder's) Seed Production

Primary seed production is done under scientific supervision at the farms of research stations, state seed farms or research and development (R & D) farms of sugar industry. Setts from well maintained seed nurseries are given heat treatment by anyone of the above detailed methods. After treatment, the setts are soaked in a fungicide solution (0.1 per cent MBC) for 5 to 10 minutes and planted in a well–prepared field, where sugarcane was not grown during the previous year. All recommended agronomical practices are followed (Fig. 9.2).

The field should be well–prepared and organic manure such as farm yard manure or cured pressmud should be applied at the rate of 25 to 30 t/ha 15 days before planting. A spacing of 75 cm to 90 cm between rows is recommended. A slightly higher seed rate of 75,000 two–bud setts is recommended for raising breeder's seed (primary

Fig. 9.2: Sugarcane planting in field

seed) nursery to compensate for germination loss due to heat therapy. For foundation and certified seed nurseries, a seed rate of 60,000 two–bud setts is adequate for obtaining a good stand.

Early application of a higher dose of fertilizer comprising 250–300 kg nitrogen, 75 kg phosphorus and 75 kg potash/ha is recommended for achieving rapid growth. The fertilizer may be applied in three split doses at 30 days, 60 days and 90 days after planting. Irrigation once in a week for loamy soil and every 10 to 12 days for heavy clay soil is adequate.

Monitoring of the seed nursery is done at least three times during the crop growth. First inspection is done at 45 to 60 days after planting to detect off–types and to remove plants infected with designated diseases and pests. The second inspection is done at 120–130 days after planting to check for off–types, designated diseases and pests. The third inspection is done 15 days prior to harvesting of canes to check the general condition of the canes as seed. The crop is harvested at 7 to 10 months and used for planting foundation seed (secondary seed) nursery. The multiplication rate is around 1:6 to 1:7, lower than the normal multiplication rate of 1:7 to 1:8 due to slightly lower germination as a result of heat treatment of setts.

9.2.2 Secondary (Foundation) Seed Production

Setts from primary seed nursery are used for planting secondary seed nursery. The crop is harvested at 7 to 10 months age and setts are used for planting commercial seed nurseries.

9.2.3 Commercial (Certified) Seed Production

Setts obtained from foundation seed nursery are used for planting commercial seed nurseries. Commercial seed plots are laid in farmers fields identified for the purpose and distributed throughout the operational area of the sugar mill. This practice avoids transport of bulky seed to long distances. The crop is harvested at 7 to 10 months age and the cane is supplied as commercial (certified) seed. Care is taken to ensure that the buds are intact during transportation. The commercial seed thus produced can be propagated for about 4 to 5 years. Seed replacement with fresh commercial seed is done only after 4 years.

The setts from commercial seed plots are supplied to the sugarcane farmers generally by the cane development department of the sugar mills. While the system of seed production and distribution works satisfactorily at some places, at several others one or more stages of the system are impaired and the seed production is affected. Thus, a large proportion of the farmers in most of the developing countries still use traditional, poor quality seeds resulting in poor cane yields (Fig. 9.3).

Fig. 9.3: Commercial seed production plots of sugarcane.

Chapter 10

Micropropagation of Sugarcane for Seed Production

Micropropagation is a methodology through which plants are multiplied rapidly by aseptic culture of meristematic regions under controlled nutritional and environmental conditions. When disease–free material is used as the source of explant or the explants are heat–treated to eliminate diseases, the resultant micropropagated plants are disease free and healthy. Micropropagation helps in the rapid seed multiplication of newly released varieties which is of particular importance in sugarcane where the normal seed multiplication rate is very low.

A number of micropropagation techniques suitable for commercial seed production in sugarcane have been reported. Apical meristem culture was used by Coleman (1970) and Hendre *et al.* (1975) to obtain sugarcane mosaic virus free plants. Axillary bud culture was applied successfully by Sauvaire and Galzy (1978) to produce true to type clones in many sugarcane varieties. Hendre *et al.* (1983) standardized an apical meristem culture technique for rapid multiplication of mosaic virus–free plants of variety Co 740. Sreenivasan and Jalaja (1981) standardized micropropagation technique based on the use of apical meristem with two or three leaf primordia (meristem tip) as the explant. The latter can be excised without the aid of a microscope and the success rate of organogenesis is quite high. The number of plantlets produced from one shoot tip in 372 days can be as high as 1,80,000. The micropropagated plants are remarkably uniform except for rare off types showing some color changes, the latter can be rouged out in the first generation itself. This meristem tip culture technique that has been widely adopted for commercial sugarcane seed production in India, is detailed in the following pages.

10.1 MERISTEM TIP CULTURE

In a growing sugarcane plant, the apical meristem is located at the tip of the stem surrounded by developing leaves and leaf sheaths.

Shoot tip explant; Shoot initiation

An established shoot culture; Shoot multiplication

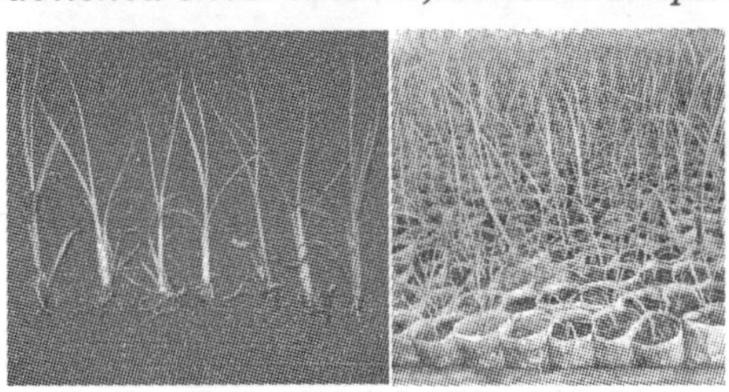

Rooted plantlets; Green house hardening

Fig. 10.1: Different stages of sugarcane micropropagation

Meristems are also located in axillary buds which are dormant as long as the apical growing point is functional. Both the apical and the axillary buds are used for initiating the shoot cultures (Fig. 10.1).

The shoot meristem measures approximately 0.1 mm in diameter and 0.25 mm to 0.30 mm in length and can be exposed by carefully removing the surrounding leaf sheaths. The meristem remains in an active state during the vegetative growth phase and the meristem cells are in a permanent embryonic state. The cells of the meristem are genetically highly stable and, hence, the plants produced from them are generally identical to the donor plants, except for the occurrence of rare mutations.

10.2 NUTRIENT MEDIA

Murashige and Skoog's (MS) medium (Murashige and Skoog, 1962) is widely employed in the micropropagation of sugarcane. MS medium without any plant growth regulator serves as the basal medium. The composition of MS medium has been given in Table 10.1.

Table 10.1: Composition of MS medium (Murashige and Skoog, 1962)

S.No.	*Constituents*	*Quantity (mg/l)*
Macro nutrients		
1.	Ammonium nitrate (NH_4NO_3)	1650
2.	Potassium nitrate (KNO_3)	1900
3.	Calcium chloride ($CaCl_2.2H_2O$)	440
4.	Magnesium sulphate, ($MgSO_4.7H_2O$)	370
5.	Pot. dihydrogen orthophosphate, (KH_2PO_4)	
6.	Sod. ethylene diaminetetraacetate ($Na_2.EDTA.2H_2O$)	37.3
7.	Ferrous sulphate ($FeSO_4.7H_2O$)	27.8
Micro nutrients		
8.	Potassium iodide (KI)	0.83
9.	Boric acid (H_3BO_3)	6.20
10.	Manganese sulphate ($MnSO_4.4H_2O$)	22.3
11.	Zinc sulphate ($ZnSO_4.7H_2O$)	8.60
12.	Sod. molybdate ($Na_2MoO_4.2H_2O$)	0.25
13.	Copper sulphate ($CuSO_4.5H_2O$)	0.025
14.	Cobalt chloride ($CoCl_{2.}6H_2O$)	0.025

Table 10.1: *(Contd...)*

Table 10.1: *(Contd...)*

S.No.	*Constituents*	*Quantity (mg/l)*
Organic constituents		
15.	Nicotinic acid	0.50
16.	Pyridoxine HCl	0.50
17.	Thiamine HCl	0.10
18.	Glycine	2.00
19.	Inositol	100
20.	Sucrose	20,000
	pH	5.8

10.3 PREPARATION OF NUTRIENT MEDIA

MS nutrient medium is prepared from the stock solutions of macro and micro salts, vitamins and hormones (Tables 10.2 and 10.3). To prepare 1 liter of MS medium, 50 ml of stock solution (1), and 5 ml of each stock solution (2), (3) and (4) are taken in a conical flask and double distilled water is added to make the final volume up to 1 liter.

Table 10.2: Preparation of stock solutions for MS medium

Constituents	*Amount (mg/l)*
MS Stock solution (1)	
Ammonium nitrate (NH_4NO_3)	33,000
Potassium nitrate (KNO_3)	38,000
Calcium chloride ($CaCl_2.2H_2O$)	8,800
Magnesium sulphate ($MgSO_4.7H_2O$)	7,400
Pot. dihydrogen orthophosphate (KH_2PO_4)	3,400
MS Stock solution (2)	
Potassium iodide (KI)	166
Boric acid (H_3BO_3)	1,240
Manganese sulphate ($MnSO_4.4H_2O$)	4,460
Sod. molybdate ($Na_2MoO_4.2H_2O$)	50
Zinc sulphate ($ZnSO_4.7H_2O$)	1,720
Copper sulphate ($CuSO_4.5H_2O$)	5
Cobalt chloride ($CoCl_2.6H_2O$)	5

Table 10.2: *(Contd...)*

Table 10.2: (*Contd...*)

Constituents	*Amount (mg/l)*
MS Stock solution (3)	
Ferrous sulphate ($FeSO_4.7H_2O$)	5,560
Sod. ethylenediaminetetraacetate ($Na_2.EDTA.2H_2O$)	7,460
MS Stock solution (4)	
Inositol	20,000
Nicotinic acid	100
Pyridoxine–HCl	100
Thiamine–HCl	20
Glycine	400

After adding sucrose and hormones in required quantities, the pH of the medium is adjusted to 5.8–6.0 with the help of a pH meter using 0.1N NaOH or 0.1 N HCl solutions. The medium (about 25 ml) is then poured in cool glass vessels duly cleaned and sterilized in hot air oven for 1 h at 160°C. The culture vessels containing the medium are plugged with aluminium foil, cotton plugs or polypropylene screw caps and sterilized by autoclaving at 15 lb pressure (121°C) for 15–20 min in an autoclave. After autoclaving, the medium is allowed to cool and used for inoculation/subculturing.

For solid medium, 0.8% Noble agar (Hi–media) is added to the warm medium and dissolved by continuous shaking and heating up

Table 10.3: Preparation of stock solutions of plant growth regulators

Plant growth regulators and abbreviations	*Weight (mg)*	*Dissolved in*	*Final volume with distilled water*	*Strength (mg/ml)*
6–Benzylaminopurine (BAP)	50	Warm 0.1 N NaOH (5 ml)	50	1.0
Kinetin (Kn)	50	–do–(5 ml)	50	1.0
α–naphthaleneacetic acid (NAA)	50	–do–(5 ml)	50	1.0
Indole–3–acetic acid (IAA)	50	–do–(5 ml)	50	1.0
Indole–3–butyric acid (IBA)	50	–do–(5 ml)	50	1.0
Gibberellic acid (GA_3)	50	Ethanol(5 ml)	50	1.0

to boiling over a heater. The warm medium is poured in glass vessels, plugged and autoclaved as described above. After autoclaving the vessels are kept stationary in a cool place for solidification.

Various concentrations of different plant growth regulators alone or in combinations are added from stock solutions (Table 10.3) to the MS basal medium in order to get specific responses (*e.g.*, establishment of shoot cultures, shoot multiplication and rooting of shoots).

10.4 PREPARATION OF EXPLANTS AND SURFACE STERILIZATION

Shoot tip explants are isolated from fresh tops collected from seed plots of desired variety of sugarcane at 8–10 month old crop age. The healthy tops are cut down and brought to the laboratory for experimentation. From the field collected tops, all the open green leaves are removed. About 8 cm long spindle segments are dissected out and washed under running tap water for about half an hour to remove the phenolics substances. The segments are shaken with mild soap solution (1% detergent powder) for 5 min and again washed thoroughly under running tap water.

The washed segments are surface sterilized with 0.1% aqueous mercuric chloride solution for 8–10 min followed by several washings with sterile distilled water. Shoot tip explants measuring about 1.0 cm in length and containing the apical dome alongwith 1–2 leaf primordia are carefully excised from the sterilized segments using sterile forceps and surgical blade and immediately used for inoculation. All the steps of sterilization should be carried out under aseptic condition.

10.5 INOCULATION OF EXPLANTS

After proper sterilization, the explants are immediately inoculated on agar solidified (8.0 g/l) MS medium under aseptic environment. All the cultures are incubated in growth chambers at 25 ± 1°C under 16 h illumination of cool white fluorescent tubes with a light intensity of about 4000 lux.

10.6 INITIATION OF SHOOT CULTURES FROM SHOOT TIP EXPLANTS

Initially the shoot tip explants inoculated on agar gelled (8.0 g/l) MS medium containing cytokinins (BAP/Kinetin) grow very slowly. Generally one or sometimes two shoots develop by the germination of pre–existing young buds. More often phenolic compounds are released from the explants which, on oxidation, produce brown coloured oxidation products that are toxic to the growing tissues. Hence, it may be required to transfer the explants on fresh medium at 1-or 2-day interval for initial 7–10 days. Shoots are developed within 15–20 days which begin to proliferate numerous tillers (side shoots) to form a clump of shoots within next 2–3 weeks. The mother shoot alongwith the explant is then carefully removed and the newly formed shoot clumps are transferred on to MS liquid multiplication medium where the new shoots grow further. These cultures are now called 'established' cultures.

10.7 SHOOT MULTIPLICATION

The established shoot cultures are transferred onto MS liquid medium supplemented with BAP, Kn and NAA (0.5 mg/l each) for shoot multiplication. Several new shoots (usually 10–25) are developed within 12–15 days to produce multiple shoots. These shoots are separated in the groups of 2 to 3 using sterile forceps and sub–cultured on fresh medium of the same composition for further multiplication. This process of separation and sub–culturing is repeated every 12 to 15 days. In this way from a single established explant, several thousands of shoots are obtained.

The zoom stereomicroscopic observations have shown that the shoot initiation absolutely takes place from the pre–existing young buds or meristems on the explants indicating their origin via axillary shoot development pathway. Mostly one or occasionally two shoots initiated from the explants. The initiated shoots grow further and gradually form a crown of tissues around its base. Most of the new shoots develop along the margin of crown through axillary shoot development pathway. The examination of shoot cultures kept under multiplication process reveal that new shoots mostly develop through axillary shoot development pathway except in some cultures where 'mixed' regeneration pathway may be followed as evident by the development of some adventitious shoot primordia. However, such structures have been observed at a very low frequency (<1.0 percent).

10.8 ROOTING OF SHOOTS

For rooting, well elongated and healthy shoots are transferred to half strength MS rooting medium containing higher concentrations of a rooting hormone (NAA, 5.0 mg/l) and sucrose (50–60 g/l). Numerous root primordia initiate from the base of shoots within 6–7 days. These primordia grow further and form well developed root system within 12–15 days of transfer onto the rooting medium.

10.9 HARDENING OF PLANTLETS

The well rooted plantlets are separated individually, the old and dried leaves are removed and the plantlets are washed under running tap water. These plantlets are kept in groups of 40–50 plants in glass bottles containing tap water and placed under growth room conditions for 4–5 days for development of secondary roots. The caps are daily removed partially for few hours. The water is also daily replaced with fresh water. After 4–5 days the plantlets were transplanted in small perforated polythene bags (7 × 10 cm) filled

Fig. 10.2: Hardening of plantlets in green house

with a mixture of soil, sand and compost (2:1:1). The plantlets are maintained in green house for 3–4 weeks at about 25–30°C under high humidity condition (about 90%). The plantlets are then shifted to shade house for 3 weeks for further acclimatization. Finally, the plantlets are transplanted to the fields (Fig. 10.2).

10.10 FIELD PLANTING OF MICROPROPAGATED PLANTS

The field in which the hardened plants are to be grown is kept fallow during the previous year to reduce soil–borne diseases. After deep ploughing, harrowing and leveling, organic manure is added @15–20 t/ha. Six meter long furrows are opened at a row–to–row distance of 90–150 cm. Pits of 10 cm depth and diameter are dug at 60–100 cm distance and a basal dose of fertilizer (75 kg P20S + 100 kg N + 40 kg K20/ha) is applied in the pits and mixed with the soil. For facilitating drip irrigation, planting is done in paired rows with 90 cm distance between rows and 180 cm distance between pairs of rows. The hardened plants are brought to the field after light trimming of the leaves and planted one per pit without disturbing the root–soil mass. The field is irrigated immediately after planting and the next irrigation is given on the third day. Subsequently, weekly irrigation is given until the first new leaf emerges which is an indication of successful establishment. First dose of fertilizer (100 kg N + 40 kg K_2O/ha) is applied at 45 days and second dose of same composition at 90 days followed by a good earthing up. Off types, if any, are rouged out at 180 to 200 days of planting.

Establishment of the tissue culture raised plants in the field is above 95 per cent if proper maintenance and irrigation schedules are followed. The crop growth is uniform with synchronous tillering and freedom from diseases and pests. The ratoons are excellent, without any gaps and ratoon yields are equal or better than the main crop yields. Canes from the ratoon crop, however, should not be used for seed production.

10.11 COMMERCIAL SEED PRODUCTION

The canes produced from the field grown, micropropagated plants are regarded as primary seed (breeder seed) under three–tier seed production system as detailed earlier. These canes are cut into two–budded or three–budded setts to raise the secondary seed (foundation seed) nursery and the seed from latter is used to raise commercial seed plots. These seed plots should preferably be located near the area where the commercial crop is to be raised, so as to minimize transportation, costs and damage during transit. A change of seed is required once in four years.

Studies carried out over several years have shown that the micropropagation–based crop has prominently better germination, tillering, cane yield, and juice content and quality than the conventionally raised crop. Heat therapy of setts also helps in improving crop and cane juice yields. The studies also revealed no detectable variations in botanical characters of micropropagation raised plants, though during the first year a few color changes in pigmented varieties and an occasional increase in tillering with a slight reduction in stalk diameter were observed. Most of the changes, however, occurred at similar frequencies as observed in the conventionally raised crop. The most common problem of excessive tillering and thin canes in the micropropagation raised seedlings (Sreenivasan and Jalaja, 1992; Sreenivasan, 1995) can be avoided by suitable modifications in the culture medium.

10.12 QUALITY CONTROL

Quality control is essential to ensure that appropriate initial material is used for micropropagation, culture conditions are satisfactory and the identity of cultivar is maintained during the culture process. The following aspects have been emphasized for maintaining the quality of tissue culture raised sugarcane plants (Sinha, 2006):

- ***Genetic Purity of Source Material:*** The genetic purity of the variety to be micropropagated should be certified by the breeder/research organization identified for the maintenance of the variety.
- ***Source Material:*** The explant should be taken from vigorously growing healthy plants raised from heat–treated setts and grown under optimum moisture and nutritional

conditions. The crop raised from micropropagated seedlings should not be used as source material.

- ***Accreditation of Micropropagation Laboratory*:** Micropropagation Laboratory should be accredited by an appropriate authority to ensure technical competence and satisfactory infrastructure,
- ***Micropropagation Protocol*:** Micropropagation protocol should ensure only minimal genetic changes. Shoot multiplication cycles should be restricted to avoid morphological variation.
- ***Seedling Establishment*:** The seedlings should be well–established in soil mixture with good root system and with 4 to 5 green leaves at the time of transplanting.
- ***Disease Indexing*:** The micropropagation–raised plants should be indexed for freedom from viruses and virus–like diseases through ELISA, and molecular methods. Standard molecular techniques may be used to assess the genetic purity of plants.
- ***Seed Production*:** The micropropagation–raised seedling should be treated as breeders (primary) seed. This seed should be further propagated through vegetative cuttings to produce foundation (secondary) seed and then commercial seed. Inspection of the field at the breeders seed production stage must be done to remove any off types.
- ***Commercial Seed*:** Commercial seed thus produced should be used up to four years.

10.13 TECHNOLOGY TRANSFER

The sugarcane micropropagation technology developed and refined at SBI during 1981–1988 was initially utilized for institute level seed production. In 1989, the technology was incorporated as a component in Sugarcane Adaptive Research Project (SARP) for implementation in nine states of India over a period of five years. The project had the objective of demonstrating that an effective seed program including the use of micropropagation technology can play a significant role in increasing sugarcane productivity. The SARP provide an effective channel to popularize sugarcane micropropagation technology, to utilize it for quality seed production on a large scale and train personnel involved in seed production (Jalaja, 1994). The program was implemented through following activities:

- A micropropagation laboratory was designed and established at SBI exclusively for large–scale clonal propagation of identified varieties. The experience gained was utilized to provide technical support for the establishment of micropropagation facilities at four locations comprising one state agricultural university and three sugar mills in two states.
- Twenty identified varieties were micropropagated and the rooted plants in polybags were supplied free of cost for seed production to 42 sugar mills located in eight states of India. The management and workers of the mills observed the superior performance of micropropagated plants. As a result, considerable enthusiasm was generated among sugar mills to establish their own micropropagation facilities, for which Government of India provided financial support.
- The SBI continued to upgrade the micropropagation technology during the subsequent years laying emphasis on reducing the production cost of micropropagated plants. The Institute started supplying cultures in flasks at multiplication stage itself to users having appropriate facilities for subsequent laboratory and field culture. This approach helped in reducing the cost and time of plant production. SBI also standardized an encapsulation technique for distribution of the micropropagated shoot initials (Jalaja, 2000).

During the implementation of SARP long–distance transportation of seedlings established in poly–bags was found to pose serious problems. The seedlings used to get damaged and the transportation charges added significantly to the cost of production. This problem was mitigated by developing, strategy of transporting plantlets without soil in plastic containers. In this procedure, the rooted plantlets from culture vessels are washed thoroughly in water and excess roots and leaves are trimmed. The plantlets are carefully packed in plastic containers keeping them erect with moist filter paper or cotton at the base and sides of the containers. In this way, 1,000 to 1,500 plantlets can be packed in one container of 10 cm diameter and 15 cm height, and several such containers can be transported in poly–bags without damage. The plantlets remain fresh for 3–4 days if transported in air–conditioned coaches. At their destination, these are planted in poly–bags, and kept in polyhouses. The usual procedures for establishment and growth are then followed. This method also helps in *in situ* hardening and good establishment and acclimatization of seedlings to the local conditions.

10.14 SUGARCANE ARTIFICIAL SEED

An artificial seed comprises meristematic tissue enclosed in a solid covering, a process called 'encapsulation'. The covering made of polymer material is permeable to air and water. The procedure of encapsulation in sugarcane involves the following steps:

i. Production of micropropagules through shoot tip culture.
ii. Separation of robust axillary shoots up to a size of 0.5 cm. Care should be taken not to damage the base of the shoots where the meristem is situated.
iii. Encapsulation of micropropagules using a 3 per cent solution of sodium alginate prepared in distilled water or in MS medium. The micropropagules are dipped in this solution and placed in 2.5 per cent calcium chloride solution for 30 min with occasional agitation. The encapsulated micropropagules can be stored up to 20 days under culture room conditions. Neelamathi *et al.* (2007) have demonstrated that these can be stored in distilled water at 1°C for 60 days with good regeneration capability.
iv. The encapsulated micropropagules can be regenerated when required by inoculating them on MS media supplemented with 1.07 mg/l kinetin and 0.5 mg/l NAA at 25°C and under illumination for 16 hrs. From this stage onwards regular micropropagation procedure is followed.

v. The encapsulated micropropagules have the advantage of easy transportability even to distantly located commercial micropropagation laboratories. However, the technology is not being used on a commercial scale since leading micropropagation laboratories are not providing this service.

Chapter 11

Tips to Reduce Cost of Cultivation in Sugarcane

11.1 SELECTION OF VARIETIES

- To give high sugar and high cane yield.
- To suit the location and season.
- Amenable for wide row spacing so as to grow intercrops.
- Tolerant to biotic and abiotic stresses.
- Possess multiratooning capability.

11.2 MAINTENANCE OF SOIL ORGANIC MATTER

- Application of green manure, greenleaf manure, compost and bulky organic crop residues to the fields under sugarcane cultivation.
- Application of well decomposed, enriched pressmud and bio–compost.
- Application of composted sugarcane trash, straw, coir pith waste, vegetable and fruit peelings, and other organic wastes.

11.3 PLANTING MATERIAL

- Good quality seed cane of 6–8 months age, free from pests and diseases.

- 75,000 two–budded setts per hectare for 90 cm spacing.

11.4 WEEDING

- Pre–emergence application of atrazine (on third day after planting) costing around Rs. 750–1,000/– can save man–days spent on manual weeding costing around Rs. 2,500/.

11.5 NUTRIENT MANAGEMENT

- Apply recommended NPK @ 260, 63 and 120 kg/ha.
- In general, soils are low in nitrogen, medium in phosphorus and high in potassium. So apply NPK fertilizers based on soil test results.
- Apply fertilizers in time, mix urea with neem cake (5:1) and place fertilizer deeply in soil.
- All the nitrogenous fertilizers should be applied within 120 days after planting.

11.5.1 Bio Fertilizers

- Apply 10 kg of phosphobacterin per hectare to save 25% phosphorous. Phosphobacterin increases the cane yield by making unavailable phosphorus into available phosphorous for the crop.
- Apply 10 kg of Azospirillum per hectare to save 25% of nitrogen.
- While applying bio–fertilizers reduce chemical fertilizers accordingly.

11.5.2 Micronutrients

- Managing the micronutrient deficiency of soils on soil test and crop symptoms of deficiency, would help to improve yields at lower cost.
- To correct iron chlorosis and zinc deficiency through repeated foliar spray of ferrous sulphate/zinc sulphate (0.5 to 2.0%) or soil application of ferrous sulphate (50 kg/ha) or zinc sulphate (25 kg/ha).

11.6 CROPPING SYSTEM APPROACH

- Avoid mono cropping of sugarcane for efficient use of nutrient and nutrients recycling.
- Follow a sugarcane production system involving ratoons, rotational crops and intercrops.
- Crop rotations with rice, leguminous crops or green manure crops would help to mobilize nutrients effectively.

11.7 USE OF LABOUR SAVING DEVICES

- Rotavator
- Cultivator
- Earthing up and ridge former
- Foot path cum channel maker
- Sett cutting machine
- Sugarcane planter
- Weeder
- Off barrer

11.8 IRRIGATION MANAGEMENT

- Deep trench planting, trash mulching along ridges saves 20.0 to 25.0% water.
- Alternate furrow irrigation saves 50% water.

11.9 PEST MANAGEMENT

- Trash mulching, frequent irrigation and light earthing up at 35th day reduces the incidence of early shoot borer.
- When the infestation exceeds 25% apply Lindane 20 EC or Chlorpyriphos 20 EC @ 5 litre/ha (5 ml per litre of water) in the whorls using a hand operated sprayer.

11.10 DISEASE MANAGEMENT

- Grow red rot resistant varieties (Co 86032, Co 86249, Co 93009 and Co 94008).

- Select disease free seed material from carefully raised nursery crop.
- Soak the setts prior to planting for 5 to 10 minutes in a solution prepared by dissolving 125 g Bavistin and 2.5 kg urea in 250 litres of water.
- Treat the setts in aerated steam therapy (AST) unit at 50°C for one hour. This treatment is required once in 5 years.

11.11 RATOON MANAGEMENT

Multiple ratooning saves:

- Cost of land preparation.
- Cost of seed material.
- Sett cutting, distribution and planting expenses.

11.12 MANAGEMENT OF HARVEST AND POST HARVEST DETERIORATION

- Harvested cane should be crushed within 24 hours.
- Post harvest cane losses can be minimized by storing under shade, sprinkling water and use of certain biocides.

11.13 LOW COST TECHNOLOGY TRANSFER

- Farmer–to–farmer interaction.
- Sharing the adaptive experience and location specific suitability with other farmers.
- Problems encountered during cultivation have to be informed to the concerned Extension Officers.

Chapter 12

Intercropping of Wheat in Autumn Planted Sugarcane

The area under sugarcane cultivation increased from 3.66 million hectares in 2004–05 to 4.13 and 4.42 million hectares in 2005–06 and 2006–07, respectively. Due to high productivity (67.4 tonnes/ha), tropical zone contributed 37.8% of India's total cane production from 33.3% area whereas, subtropical zone with 66.6% area under sugarcane contributed only to 62.2% of cane production due to lower productivity (53.3 tonnes/ha).

It is estimated that 9.69 lakh hectares of sugarcane area in the country is under Sugarcane–Sugarcane Ratoon–Wheat system. The

Fig. 12.1: Sugarcane planting in furrows in standing wheat

importance of the system lies in the fact that more than two–third of the sugarcane area in western U.P. and 10–14% in other sub–tropical states (Uttaranchal, Punjab, Haryana, Bihar, Madhya Pradesh and Rajasthan) is covered by this system. Late planting (summer planting) of sugarcane after wheat harvest (April–May) reduces the sugarcane yield by 30–50%, due to less germination under higher heat load, which results in poor stand of sugarcane.

12.1 METHODS OF PLANTING

In order to improve sugarcane productivity in wheat–sugarcane sequential system, intercropping under Furrow Irrigated Raised Bed (FIRB) system with wheat sowing on raised beds and sugarcane planting in furrows has been developed. The sugarcane + wheat intercropping has been exploited to enhance sugarcane production in this area without loosing the wheat crop by autumn planting of sugarcane during October–November. In this system sugarcane planting is advanced to October which otherwise is done in late April/ May in wheat–sugarcane sequential system.

12.2 USE OF RAISED BED PLANTER

In this system 2–3 rows of wheat are sown on each raised bed (35–40 cm top width) made by tractor drawn Raised Bed Maker–cum–Fertilizer Seed Drill (Fig. 12.2) with furrow to furrow spacing of 67–70 cm and the sugarcane setts are placed in the furrows manually.

Fig. 12.2: Machine making raised bed–furrow system

In wider spaced raised beds (65–95 cm top width) 4–5 rows of wheat are sown on beds and setts are placed manually in furrows at 90–120 cm row to row spacing for sugarcane. Immediately after placing, the setts are covered by a layer of soil.

12.3 SELECTION OF VARIETIES

Early sown varieties of wheat (WH 542, PEW 343, PEW 502, DEW 17 etc.) are suitable for sowing with sugarcane (early maturing varieties like Co 89003, CoH 56, Co 98014 etc.), as inter crop.

12.4 SOWING TIME AND SEED RATE

Optimum sowing time for wheat + sugarcane intercropping varies from 25 October to 7 November in sub–tropical region keeping the seed rate of wheat at 75–80 kg per hectare. For planting one hectare area, 87,500 two budded setts or 57,500 three budded setts are required, which can be obtained from 8.5–10.0 tonnes of sugarcane (Fig. 12.3).

Fig. 12.3: Sugarcane stands after wheat harvest

12.5 SYSTEM BASED NUTRIENT MANAGEMENT

- Fertilizer is to be applied as per the individual crop requirement.

- For wheat, optimum dose of N: P: K is 150:60:40 kg/ha 1/3 N + full P and K at sowing, 1/3 N at 20–25 days after sowing (DAS) and remaining 1/3 N at 45–50 DAS are applied.
- After wheat harvest, full dose of P and K (50:50 kg/ha), 1/3 N (50 kg/ha) and 2 litres/ha chlorpyriphos to control termites are applied in furrows 3–4 days after irrigation.
- Immediately after application of fertilizers, one hoeing with partial earthing up of soil should be done so that the applied nutrients are covered by soil which will enhance the nutrient use efficiency.
- 1/3 N (50 kg/ha) is applied at May end and remaining 1/3 N is applied at initiation of grand growth stage or at onset of monsoon in June end.

12.6 IRRIGATION

- Immediately after covering the sugarcane setts in furrows with soil, a light irrigation filling 3/4th of the furrow is required for proper germination of wheat and sugarcane.
- Second light irrigation is to be applied 8–10 days after sowing and subsequent irrigations are given in the furrows as per crop requirement and prevailing weather conditions.
- After wheat harvest, the furrows are used for irrigating sugarcane till earthing up operation.
- Before the onset of monsoon, irrigations are given at 10–12 days interval whereas, 20–25 days interval may be sufficient during the monsoon period as per crop requirements.

12.7 WEED MANAGEMENT

- To manage both broad leaf as well as grassy weeds, one intercultural operation should be carried out with hand hoe at 25–30 days.
- If weed population is more, it may be effectively controlled by application of a suitable herbicides.
- Spraying of Carfentrazone @ 20 g a.i./ha and Fenoxaprop @ 100 g a.i./ha in 250–300 liters water at 30–35 DAS is effective to control broad leaved weeds and grassy weeds, respectively.

- Isoproturon @ 1000 g a.i./ha or Sulfosulfuron + metsulfuron @25 = 4 g a.i./ha applied at 30–35 DAS in 250–300 liters water is effective to control complex weed floor.
- After harvesting of wheat, '1–2 intercellular and earthing up operations in sugarcane will take care of remaining weeds.

12.8 HARVESTING OF INTERCROP

- Wheat should be harvested at physiological maturity and removed from the field to threshing floor. Irrigation and fertilizers should be applied to the sugarcane crop immediately after wheat harvest.

12.9 BENEFITS OF INTERCROPPING

- Sugarcane planted in the autumn (October–November) completes its germination and to some extent tillering phase by the time of wheat harvest. Otherwise, it is delayed by nearly 2–3 months to spring planting (February–March) or 4–5 months to summer planting (April–May) in wheat–sugarcane sequential system.
- Autumn planted sugarcane + wheat results in higher cane yield without loosing wheat yield of the system and better sugar recovery with more gross returns.
- It leads to better utilization of land, water and sun light. The system increases water use efficiency by saving irrigation on water (20%) as it is applied only in the furrows requiring less volume of water to wheat as compared to flat method. Moreover, no extra irrigation is needed for field preparation and establishment of sugarcane till the harvest of wheat.
- The system serves as an alternate weed management technology because lower weed counts are observed on raised beds and weeds mostly colonize in the furrows, which are moist. Weeds could be controlled easily by mechanical weeding or directed herbicidal spray.
- The system holds promise to increase input use efficiency and system productivity.

- The technology reduces production cost and increases margin of profit with efficient use of resources.
- Generates additional employment for the farm family or agricultural labourers.

Cropping pattern	*Sugarcane equivalent productivity (tonnes/ha)*	*Net returns (Rs./ha)*	*Input output ratio*
Sugarcane–ratoon–wheat	125.7	51945	2.74
Sugarcane+wheat–ratoon	160.2	75845	3.68

12.10 RELATIVE PERFORMANCE OF SEQUENTIAL AND SIMULTANEOUS PLANTING OF SUGARCANE–WHEAT

Under present conditions, the chances of increasing production by increasing the total cane area at the cost of other field crops are limited. Hence simultaneous planting of wheat in autumn planted sugarcane (intercropping) offer greater scope for increasing production and productivity by enhancing resource use efficiency and intensification of cropping system.

Chapter 13

Package of Practices for Sugarcane Cultivation

Potential productivity of sugarcane is over 300 t/ha, but the average yield obtained in the country is only around 70 t/ha. Scientific management practices for sugarcane are detailed below:

Season: Best time to plant sugarcane in the tropical states is from January to March. The adsali crop is planted from July to September.

Varieties: A variety combining high yield, quality and ratooning potential needs to be chosen for specific locations. Currently Co 86032 is the most suitable and popular variety in the tropical belt.

Land preparation: A green manure crop (dhaincha or sesbania) has to be raised prior to sugarcane and incorporated at following stage by a disc plough or a rotavator. After about two weeks, further deep ploughing once or twice needs to be done followed by harrowing to bring the soil to good tilth. In the absence of green manuring, farm yard manure or compost (25 t/ha) need to be applied before last ploughing in the furrows.

Spacing: Ridges and furrows are formed at 90 cm. Wider spacing of 120 cm may be followed in fertile soils, for high tillering varieties, and where irrigation water is in plenty, the furrows should be made 30 cm deep.

***Seeds*:** Sugarcane setts should be obtained from nursery crops of around 8 months age. Two bud setts are preferable. Setts should have healthy buds and free of diseases and pests. About 75,000 two bud setts/ha is sufficient. Setts should be soaked for 10 minutes in a solution prepared by dissolving 125 g Bavistin and 2.5 kg urea in 250 liters of water.

***Basal manuring*:** 10% of total recommended nitrogen (30 kg/ha) and the entire dosage of phosphorus (63 kg P_2O_5/ha) should be applied in the furrows and mixed with soil.

***Planting*:** Setts are planted end–to–end in the furrows, buds facing the sides and covered with soil in dry planting, in light soils. In heavy soils, the furrows are irrigated and setts are pressed into soil.

***Weed control*:** On 3rd or 4th day after planting, atrazine is sprayed @ 2 kg a.i./ha using a suitable sprayer. This is followed by hand hoeing and weeding around 40–45 days (before first top dressing). If broad leaved weeds are present, a post–emergence spray of 2,4–D @ 1.0 kg a.i./ha may be given after one month of planting. Before second top dressing, hand weeding may be done. A directed spray of atrazine after final manuring and earthing up would help to control late weeds.

***Top dressing*:** 45% of the nitrogen dosage (125 kg/ha) and 50% of potassium (60 kg K_2O/ha) are applied at 45 days after planting. Partial earthing up is done. To improve the efficacy of urea, it may be blended with neem cake (urea: neem cake = 5:1) and pocket manuring technology may be adopted. The second top dressing is done at 90 days after planting with the remaining 45% of nitrogen (125 kg/ha) and 50% of potassium (60 kg K_2O/ha) and full earthing up is given.

***Biofertilizers*:** By applying *Azospirillum* and Phosphobacteria each at 10 kg/ha, 25% economy in the chemical fertilizer can be made. Biofertilizers are applied on 30 and 60 days after planting @ 5 kg each, each time, by mixing with powdered farm yard manure.

***Irrigation*:** The following irrigation schedule may be followed:

Germination (0–35 days)	Weekly
Tillering (36–100 days)	Once in 10 days
Grand growth (101–270 days)	Weekly
Maturity (271–harvest)	Once in 15 days

This schedule is for medium type soils. In heavy soils, 10–15 days interval up to grand growth stage, and 15–30 days thereafter.

***Detrashing*:** Dry and senescent leaves are stripped at 5th and 7th month and may be applied as mulch in alternate furrows.

***Prevention of lodging*:** To prevent lodging, heavy wet earthing up at 7th month and trash–twist propping may be followed.

***Plant protection*:** As a preventive measure, healthy seeds should be used. Constant vigil is to be maintained on the occurrence of any pest or disease and plant protection measure should be need based, with the help of agricultural officers, sugar factory and cane department.

***Harvest*:** Harvesting should be done at full maturity at 12 months. Close harvesting (2–3 cm below ground level) using a hand axe is ideal. Topping should be done at the point of break. Clean cane should be supplied to the mill or used for jaggery making. The cane should be transported to the mill within 24 hours.

***Cane yield*:** With the package of practices detailed above, a cane yield of 150–200 t/ha can be obtained.

***Ratooning*:** Within a week of harvest, ratooning operations such as stubble shaving and off–barring may be carried out to obtain good ratoons. Gap filling and early manuring are also essential.

Chapter 14

Sugarcane Ratoon Management

Raising productive ratoons is essential for over–all profitability. By ratooning, about 30% economy in the operational cost could be achieved through saving seeds, labour and irrigation besides time. At present, ratoon yields are generally lower by 20–30 percent when compared with plant crop. But by scientific ratoon management practices as detailed below ratoon yield could be improved to the level of plant crop.

14.1 CHOOSING A VARIETY WITH GOOD RATOONING POTENTIAL

Varieties differ in their ratooning ability. Varieties having high tillering ability, building higher stalk population, medium in girth and high plant crop yield are generally good ratooners (Fig. 14.1).

14.2 GOOD MANAGEMENT OF THE PLANT CROP

Plant crops raised with all the essential management practices, harvested at appropriate time/age, having adequate stalk population at harvest and free of carry over pests and diseases will give good ratoons. Crops harvested during extreme cold or hot weather period do not give good ratoons and need special management practices.

Fig. 14.1: Ratoon

14.3 STUBBLE SHAVING

Stubbles of the plant crop (previous crop) should be cut below the soil level to about 5 cm with a sharp blade, hand axe or stubble shaving machine to promote healthy underground buds to sprout and give a good initial crop stand with better root system.

14.4 OFF–BARRING/SHOULDER BREAKING

The sides of the ridges are cut using a spade or by passing a narrow ridger or plough to loosen the soil and to improve aeration. This operation is also known as root pruning.

14.5 GAP FILLING

In any row, if for a distance of 60 cm, no sprouting is seen, it needs to be filled. For gap–filling, one month old polybag seedlings are ideal. Seedlings are placed in the gap after clipping the leaves and pot watered till estabilishment (Fig. 14.2).

14.6 EARLY MANURING

Early manuring as per the schedule below is essential. The dosage could be same as plant crop or 25% extra N could be given:

At ratooning: Full P + 1/3 K (469 kg super phosphate + 203 kg urea + 67 kg muriate of potash)

- 30–40 days – 1/3 N + 1/3 K (203 kg urea + 67 kg muriate of potash)

Fig. 14.2: Polybag seedling for gap filling

- 60–70 days – 1/3 N + 1/3 K (203 kg urea + 67 kg muriate of potash)

Fertilizers are applied as band placement on the sides after opening shallow furrows and covered. Pocket manuring (placing the fertilizer in 10 cm deep holes made by a crow bar 10 cm away from the clumps) gives better result. Organic manures at 25 t/ha may be applied in addition to chemical fertilizers.

14.7 CONTROL OF CHLOROSIS

Repeated application of ferrous sulphate at weekly intervals at 0.25 to 0.5% concentration would help overcome chlorosis.

14.8 WEED CONTROL

After stubble shaving and off–barring, direct application of atrazine at 1.5 kg a.i./ha on the interspaces will help control weeds for 30–45 days. Later, post emergence application of 2, 4–D @ 1.0 kg a.i./ha would help control broad leaved weeds. *In situ* trash mulching helps in checking weeds.

14.9 IRRIGATION

Frequent light irrigations are essential to facilitate early sprouting. Later, irrigations could be given once in 7–10 days in light and medium soils, and 12–15 days in heavy soils.

14.10 PESTS AND DISEASE MANAGEMENT

In general ratoons are less vigorous and hence more prone to attacks by pests and diseases. Three diseases *viz.*, ratoon stunting disease (RSD), grassy shoot disease (GSD), and smut are particularly more common in ratoons. To avoid these diseases, a healthy nursery programme involving heat therapy is necessary. RSD and GSD can be overcome by any of the heat treatment practices (aerated steam or moist hot air). For smut, hot water treatment with a fungicide would be effective. Heat treatment and production of healthy primary seed should be responsibility of the sugar factories or research stations or any other seed producing agency. Farmers should obtain seeds from such nurseries. Farmers can also get the primary or secondary seeds and then produce seeds for their own use. GSD affected clumps which could be easily identified should be uprooted and destroyed. Likewise, smut whips should be removed before the whip ruptures and then burnt.

14.11 *IN SITU* TRASH MULCHING AND INCORPORATION

Sugarcane trash available after the harvest of the crop, could be used for mulching which conserves soil moisture in the fields (Fig. 14.3).

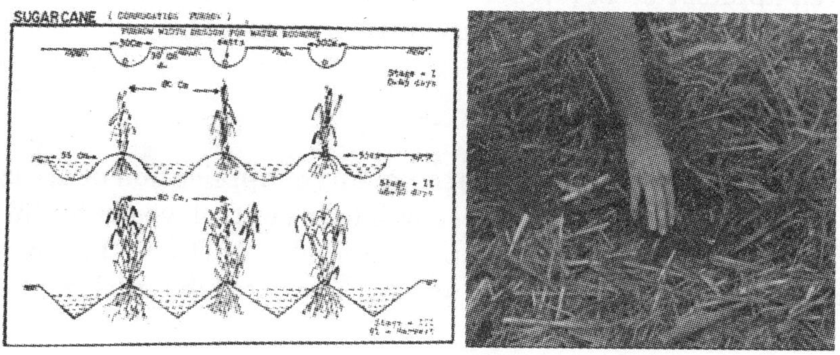

Fig. 14.3: Sugarcane trash for Mulching

Chapter 15

Integrated Nutrient Management in Sugarcane

Sugarcane is a long duration (10–14 months) crop and produces a large quantity of biomass. On an average, one tonne of sugarcane crop removes 1.0, 0.6 and 2.25 kg N, P_20_5 and K_20 from the soil. Thus, to produce 100 tonnes/ha cane yield, it removes 100, 60 and 225 kg/ha N, P_20_5 and K_20 and a large amount of micronutrients from the soil. The results of experiments on manures and fertilizers have revealed that neither the fertilizers alone nor the organic sources exclusively can achieve sustainability of sugarcane production and maintain soil health under highly intensive cropping systems being practiced today. Therefore, integrated nutrient management (INM), involving the use of fertilizers in conjunction with organic sources and bio–fertilizers is essential to achieve higher productivity of sugarcane.

The basic principle of INM is the maintenance of soil fertility, sustainable productivity and improving farmer's profitability through judicious and efficient use of organic manures, chemical fertilizers, green manures, bio fertilizers, industrial by–products etc.

15.1 FERTILIZERS

Fertilizer is the most important component of INM for sugarcane production and efficient fertilizer use is a primary component of a farmer's best management practice. Full dose of P and K (Table 1) should be placed in the furrows before planting to overcome the problem of low mobility of P. Application of N in three equal splits *i.e.* 1/3 at planting, 1/3 at the tillering phase (60–90 days after planting) and remaining 1/3 N at initiation of grand growth stage (100–115 days after planting) before June end is found to be best manurial practice. Immediately after application of fertilizers, one hoeing with partial earthing up of soil should be done to cover the applied nutrients, which minimizes the loss of nutrients from the soil and improves fertilizers use efficiency.

Table 15.1: Recommended dose of fertilizer for sugarcane in different sub tropical states

Name of state	***Recommended dose of fertilizer (kg/ha)***		
	N	P_2O_5	K_2O
Bihar	90–120	60–70	30
Haryana	150	50	–
Madhya Pradesh	320	80	45
Punjab	150–180	60	–
Rajasthan	150	50	–
Uttarakhand	120–150	60–80	40
Uttar Pradesh	150–180	60–80	40

15.2 LEGUMES AND GREEN MANURE

Legumes in sugarcane based cropping system are grown either in sequence or as intercrops for green manuring/grain/fodder. Green manures (dhaincha and sunhemp) or pulses (black gram, green gram, cluster bean, cowpea etc.) preceding sugarcane give a benefit of 19–43 percent increase in the yield of sugarcane and contribute 30–85 kg/ha nitrogen. Integrated use of green manure crops along with mineral fertilizers supplements the major nutrients to sugarcane besides increasing availability of micronutrients and improves nutrient use efficiency. Growing of legumes in inter row spaces as intercrops for grain/fodder/green manuring in sugarcane and incorporation of biomass/crop residue is a very useful practice to economise the cost of cultivation (Fig. 15.1).

Fig. 15.1: Green manure

15.3 ORGANIC MANURES

The organic manures release nutrients slowly over a longer period and play an indirect role by improving the physico–chemical and biological properties of the soil. Long term field studies on sugarcane have shown the need for applying organic manures (FYM, compost, press mud, trash etc.) as a basal dose for improving the soil fertility. With increasing deterioration of soil health, there is an urgent need to reintroduce the practice of organics application. Application of 10–15 tonnes/ha of FYM at the time of field preparation is adequate to meet 20–25% of nutrient requirement of sugarcane (Fig. 15.2).

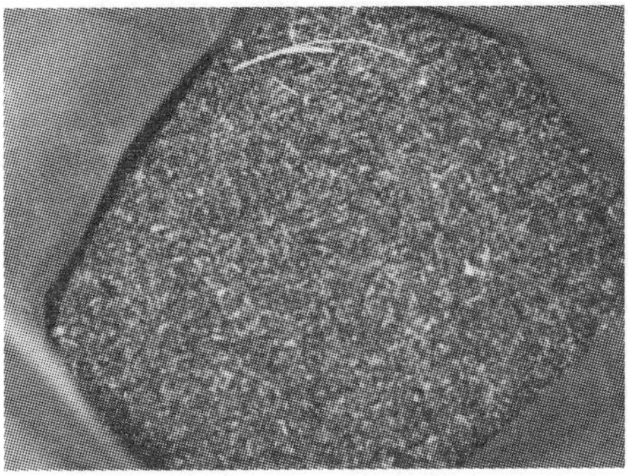

Fig. 15.2: Sugarcane ash as organic fertilizer

15.4 CROP RESIDUES AND TRASH

Addition of sugarcane trash compost has an effect similar to that of a green manure crop incorporated into the soil. About 15–20 tonnes of trash, which contains about 0.35% N, 0.13% P_2O_5 and 0.65% K_2O along with appreciable amounts of micronutrients, are produced in one hectare area. But it has a wide C:N ratio. Therefore, trash should be used for trash mulching or making nutrient rich compost and utilized as organic manure. The positive value of trash mulching lies in its capacity to minimize N loss from the soil and its incorporation after partial decomposition increases organic matter level in the soil. Sugarcane trash can be easily composted using fungal cultures *(Pleurotus or Trichoderma or Aspergillus)*, pressmud, cow–dung as starter and a mixture of rock phosphate, gypsum and urea in the ratio of 5:4:1. An integrated use of 5 tonne/ha cane trash with 100 kg/ha N in sub–tropics economises nearly 1/3 of N applied to sugarcane.

15.5 FACTORY WASTES

Filter cake or pressmud cake, a byproduct of sugar industry, is produced @ 3–7 percent of the cane, crushed. On an average, each tonne of sulphitation pressmud cake (SPMC) contains 12.6, 38.3, 14.6 and 23 kg of N, P, K and S respectively besides micronutrients. Fresh pressmud has wide C: N ratio and releases a lot of heat during decomposition. Hence, it is advised that pressmud should be applied only after proper composting. To convert every tonne of pressmud to compost, add microbial cultures (*Pleurotus* or *Trichoderma viride*) @ 1 kg, urea 5 kg and cow dung or FYM @ 50 kg as a starter and sprinkle by mixing them in water on the layers of pressmud, spread to a thickness of 30 cm. This process is repeated to a height of about one metre. Then water is sprinkled periodically to maintain moisture at 50% water holding capacity. Decomposition will be over in 6 to 8 weeks. Application of 10 tonnes/ha of composted and enriched press mud could save 25% of chemical fertilizers and enhance N use efficiency by 4–8% in sugarcane.

15.6 BIO–FERTILIZERS

Bio–fertilizers (Fig. 15.3) play an important role in supplementing the nutrient requirements through nitrogen fixation and phosphate solubilization. Nitrogen fixing bacteria like *Azotobacter,*

Azospirillium and *Acetobacter* and phosphorus solubilising bacteria *(Bacillus megaterium)* have been recognized as efficient bio fertilizers for sugarcane and found to economise fertilizer N and P by 20–25 percent. For soil application, bio fertilizers (10 kg *Azospirillium* + 10 kg phosphorus solubilising bacteria (PSB) or 10 kg *Azotobacter* + 10 kg PSB or 10 kg *Acetobacter* alone per hectare) are mixed thoroughly with 500 kg of finely powdered FYM/compost/ pressmud and applied in two splits at 35–45 and 65–75 days after planting near the base of the clumps followed by partial earthing up and irrigation to ensure optimum moisture conditions for bacterial growth. Biofertilizers can also be applied by dissolving in water and applying near to the clumps after 2–3 days of irrigation.

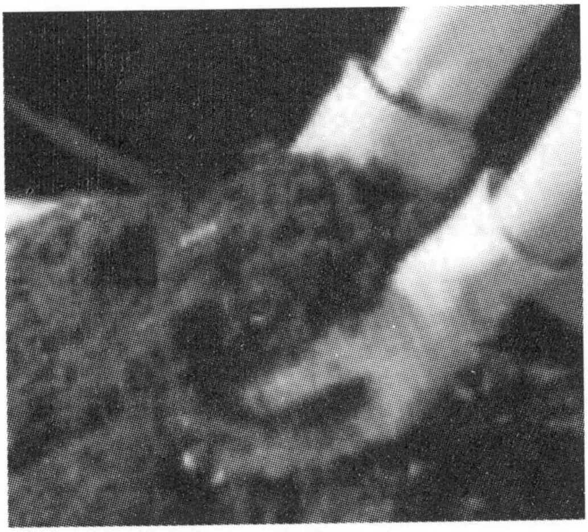

Fig. 15.3: Biofertilizer

Care should be taken that there should be minimum of one week gap between application of biofertilizers and insecticide/weedicide/ fertilizer applications.

A suitable combination of chemical fertilizers (Fig. 15.4), organic sources, green manures and bio fertilizers is essential to harvest a good cane crop without deteriorating soil health for long term sustainability.

Fig. 15.4: Chemical fertilizer

Chapter 16

Biofertilizers for Sugarcane

Biofertilizers are preparations containing live or latent cells of efficient strains of nitrogen fixing, phosphate solubilizing or cellulolytic microorganisms used for application to seed, soil or composting areas with the objective of increasing the population of specific microorganism and to accelerate certain microbial process to improve the availability of nutrients (Fig. 16.1).

Nutrient requirement of sugarcane is high and at present this is mostly met by the application of chemical fertilizers. Integrated nutrient management is important to stabilize and sustain yields of cane and sugar. Biofertilizers provide one or more nutrients, increase

Fig. 16.1: Biofertilizer

availability and improve soil physical, chemical and biological properties. Hence, biofertilizers should be an important component of integrated nutrient management practice.

16.1 NITROGEN FIXING BIOFERTILIZERS

Nitrogen fixing biofertilizers contain efficient microorganisms capable of fixing elemental nitrogen into plant usable form. In sugarcane, three bacteria *viz.*, *Azotobacter, Azospirillum* and *Acetobacter* are used as nitrogen fixing biofertilizers.

1. *Azotobacter* is an aerobic and free living bacteria found in high numbers in the rhizosphere. *Azotobacter* uses soil organic matter and plant root exudates for fixing atmospheric nitrogen and produces growth promoting substances like Gibberellic acid, Indoleacetic acids and polysaccharides that help in improving the physical characteristics of soil. They are relatively more effective in the soils rich in organic matter.
2. *Azospirillum* is an efficient nitrogen fixing bacteria capable of colonizing the roots of plants. The low oxygen availability conditions prevailing in rhizosphere and root surface is ideal for its growth and nitrogen fixation. It also produces growth promoting substances and polysaccharides.
3. *Gluconacetobacter diazotrophicus (syn. Acetobacter diazotrophicus)* occurs mainly in sugarcane and also in crops like sweet sorghum, coffee and sweet potato. This bacterium is an endophyte, which can colonize the root surface and the inner parts of the sugarcane plant. This bacterium fixes large quantities of atmospheric nitrogen and efficiently transfers the fixed nitrogen to sugarcane.

A large number of field trials in Tamil Nadu revealed that application of *Azospirillum* biofertilizer is more suitable than *Gluconacetobacter. Azospirillum* application increased the sugarcane yield by 10 per cent with 25 per cent saving in N fertilizer besides improving the juice quality. Application of biofertilizer with lower dose of N fertilizer is more beneficial than with the recommended N dose.

16.2 PHOSPHATE SOLUBILIZING BIOFERTILIZERS

Phosphorus is a key nutrient for higher sugarcane and sugar yields. But in most soils, availability of phosphorus is extremely limited due to fixation *i.e.,* conversion of available phosphorus to unavailable or difficultly available form. Only 10 to 15 per cent P is available to the crop to which it is applied. Many soil bacteria and fungi convert this insoluble inorganic phosphate into soluble forms through production of organic acids. These, microorganisms are known as phosphate solubilizing microorganisms. The efficient phosphate solubilizing microorganisms are *Bacillus megaterium; Pseudomonas striata* and *Aspergillus awamori.* Among these, *Bacillus* megaierium var. *phosphaticum* (phosphate solubilizing bacterium, PSB) has been found suitable for sugarcane. Field experiments have indicated 10 to 15 per cent increase in cane yield. There was also improvement in the juice quality and available P status of the soil. When phospho-bacteria is applied, costly super phosphate can be substituted with cheap rock phosphate up to an extent or 50 percent on nutrient basis.

16.3 QUALITY OF BIOFERTILIZER

To produce biofertilizers, the microorganisms concerned are grown in suitable nutrient medium and mixed with stabilized carrier materials, like peat or lignite. This biofertilizer material should contain about 25 to 35 per cent moisture and sufficient microbial population. As per BIS standards the biofertilizer should contain 1 $\times 10^8$ colony forming units (population) per gram and 1×10^7 colony forming units per gram at expiry.

16.3.1 Dosage

The recommended dose of biofertilizer is 10 kg *Azospirillum* and 10 kg *Phosphobacteria* per hectare for soil application and 2 kg *Gluconacetobacter* for sett dipping.

16.3.2 Method of Application

For sett dipping, mix 2 kg of biofertilizer in 500 litres of water, soak the setts for 30 minutes and plant immediately.

For soil application, apply the recommended dose of biofertilizers in two splits. For fields wherein chemical fertilizers are to be applied in two splits (45 and 90 days), apply biofertilizers on 30 and 60 days, at the rate of 5kg *Azospirillum* and 5 kg phosphobacteria each time. For fields wherein chemical fertilizers are applied in three splits (30, 60 and 90 days), apply biofertilizers on 45 and 75 days, at the rate of 5 kg *Azospirillum* and 5 kg phosphobacteria each time. *Mix Azospirillum* and phosphobacteria thoroughly with about 500 kg of powdered farmyard manure and apply uniformly near the base of the sugarcane clumps. Immediately give a light earthing up and irrigate.

16.4 OVERALL BENEFITS

- 25% saving of nitrogenous and phosphatic fertilizers.
- Possibility of 50% replacement of super phosphate with rock phosphate.
- Better crop growth leading to around 10% increase in cane yield.
- Improvement in soil health and quality.

***General Precautions*:** The biofertilizers are only supplementary to chemical fertilizers to meet the nutrient demand of the crop. They supplement nitrogen and improve the scope for using cheap low–grade P sources. Give 10–15 days gap between the applications of biofertilizers and chemical fertilizers. Efficacy of biofertilizers can be improved by the addition of organic manures. Maintain sufficient soil moisture to obtain best results.

Do not mix biofertilizers directly with chemical fertilizers, pesticides, herbicides or fungicides. Store the biofertilizer packets at room temperature away from direct sunlight. Biofertilizers should be obtained from a reliable source.

Chapter 17

Sugarcane Management in Saline Soils

Sugarcane is grown in 4.2 million hectares in India and about one fourth of the acreage is affected by salinity and/or alkalinity to varying degrees. Soil salinity and poor quality irrigation water coupled with moisture stress during high water demand period (in summer months), largely coinciding with rapid growth phase of sugarcane results in very low yields. It is estimated that 33% of cane area in Tamil Nadu, 40% in Andhra Pradesh and 48% in Karnataka experience either soil salinity or saline migration water and yield losses reported were about 40 per cent.

17.1 SALTS CONTRIBUTING TO SALINITY

Chlorides and sulphates of sodium, calcium, magnesium and potassium largely contribute to salinity of the soil while sodium carbonates and bicarbonates contribute to alkalinity.

17.2 RESPONSE OF SUGARCANE TO SALINITY

General response of sugarcane crop to high salty soil is the poor and delayed germination, reduced tillering, leaf yellowing and burning, stunted growth, poor crop stand, extended growth and reduced yields. The salt interferes with sugar production by affecting growth rate and cane yield and by reducing the sucrose content of stalk. Salt increases the fibre content of the cane and electrical conductivity (EC) of juice, which affects the jaggery preparation and jaggery quality.

17.3 MANAGEMENT PRACTICES

An integrated approach involving the measures indicated below should be employed to manage sugarcane under salinity and to improve productivity:

(i) ***Seed Rate***: Higher seed rate of 25% is recommended to compensate for germination loss and to ensure adequate crop stand.

(ii) ***Trench Planting***: Following "modified trench system" of planting in saline water irrigated areas, improved yields of around 15 per cent has been recorded. In this method, while earthing up, furrows are not converted into ridges; instead a trough is maintained along the row. The irrigation water is let in the cane row itself.

(iii) ***Use of Organic Manure***: Organic manures *viz*, pressmud (10–15 t/ha), farmyard manure (25 t/ha), bioearth etc. improve the availability of essential nutrients (Zn, Fe, Ca, Mg and Mn). In calcareous soil the organic manures reduce the soil pH, electrical conductivity (EC) and exchangeable sodium per cent, rendering the soil more suitable for growing sugarcane

(iv) ***Amendments***: With increase in soil pH the requirement of gypsum also increases. However, for most soils gypsum at 3–6 t/ha would be sufficient.

(v) ***Irrigation with Good Quality Water***: During critical growth stages (upto 150 days of crop age) irrigation with good quality water is beneficial.

Mulching: Trash mulching reduces loss of moisture through evaporation there by minimizing the effect of ions in the soil. Trash upon incorporation adds organic matter and nutrients (Fig. 17.1).

Fig. 17.1: Sugarcane mulching

(i) ***Green Manures***: Growing green manure intercrop and *in situ* incorporation of green manure is highly beneficial to improve productivity in salt affected soils.

(ii) ***Nutrient Management***: 25% additional nitrogen dosage has been found to improve yields under saline conditions. Application of top dressings of nitrogen and potassium fertilizers through pocket manuring is advantageous and helps in improving yields significantly. For this, small holes of about 10 cm deep and 10 cm away for the clump are made using a crow bar, fertilizers are placed and covered.

(iii) ***Crop Rotation***: Crop rotation with tolerant crops like cotton, mustard etc.

(iv) ***Tolerant Varieties***: Growing salt tolerant varieties indicated below.

17.4 SUGARCANE GENOTYPES TOLERANT TO SALINITY

Varieties Co 6806, Co 7219, Co 7717, Co 8208, Co 85004, Co 85019, Co 86032 and CoC 671 are suitable for salt affected soils. New tolerant genotypes have also been identified (Table 1).

Sugarcane yields under saline soils could be substantially improved by following the management practices recommended as above.

Table 17.1: Promising genotypes possessing tolerance to salinity

Co 87002	Co 89010	Co 89027	Co 90010	Co 91002
Co 91005	Co 91011	Co 92012	Co 93005	Co 93009
Co 93011	Co 93015	Co 93016	Co 93018	Co 93019
Co 94004	Co 93021	Co 93079	Co 94005	Co 94008
Co 94010	Co 94011	Co 94012	Co 94015	Coln 86141
Coln 94141	CoG 93076	Co 95003	Co 95006	MS 92121

Chapter 18

Sugarcane under Salt Affected Conditions

Sugarcane is one of the major crops of our country, cultivated in almost all the states. Of the 3.66 million ha of land under sugarcane during the 2004–05 crop season, nearly one fourth is salt affected. Irrigation with poor quality water, coupled with water deficit conditions in the salt affected soils during summer month, leads to severe reduction in crop growth which ultimately leads to lesser sugar production under these conditions.

18.1 HOW DO THE SALT AFFECTED FIELDS APPEAR?

Salt affected soils may be saline and/or alkaline in nature. The problem is often obvious as the fallow fields have a shinning layer of white crystals over their surface in highly saline areas whereas at lower levels of salinity, patchy salt crusts appear only on the top portions of the ridges, raised areas of unleveled fields or under the soil clods. The surface of only sodic/alkali soil appears to be black and deep fissures appear when these soils dry up. However, saline–sodic soils appear to be as if they are saline ones.

18.2 WHAT MAKES THE SOIL OR WATER SALT AFFECTED?

High concentrations of chlorides and sulphates of sodium, calcium and magnesium in various proportions make a soil saline. If high sodium ions are associated with carbonates and bicarbonates, then the soil becomes sodic over a period of time as carbonates of calcium are primarily insoluble which shifts more and more of calcium from the exchange sites into the soil solution in exchange for sodium. A soil which has more than 15 ESP (exchangeable sodium percent) is termed as sodic and generally has a pH above 8.2. A soil is termed saline if its E.C. is more than 4 dSm^{-1} and saline sodic if both the conditions are met with.

18.3 EFFECT OF SALINITY ON SUGARCANE PLANTS

- Salinity slows down and decreases germination of sugarcane setts. In the life of a sugarcane plant, the period following bud sprouting till the plant develops at least 2 fully expanded leaves is the most sensitive period to salinity.
- Reduced growth with stunted plant appearance, poor germination and tillering gives the sugarcane fields a patchy look leading to reduced yields.
- Accumulation of excess of Na^+ and Cl^- ions brings about yellowing followed by necrosis of margins of leaves which starts from the tip region and proceeds down–wards and inwards. These symptoms arise first on the oldest leaf to proceed upward towards the younger ones. Thus the old leaves senesce earlier thereby decreasing the rates of growth.
- The sugarcane plants growing under salt stress conditions start shinning with the first monsoon rains as plants pick–up growth immediately. Measuring leaf elongation rates with a ruler has shown them to be as good as that of the normal growing plants within 2–3 hours of desalinization.
- Salt stress not only reduces the rate of photosynthesis but also affects its utilization in growth.
- Medium level of salinity normally reduces the level of sucrose in juice which results from lower level of transport of sugars from the leaves to the internodes. Response may vary with the variety being grown.

- High levels of salinity increase the fibre content of canes and excess of soluble salts in the juice make jaggery making, a difficult process and also result in poor quality jaggery.

18.4 HOW TO GET A GOOD SUGARCANE CROP UNDER SALINE CONDITIONS?

- Use 25% more seed cane in order to compensate for reduced NMCs at harvest which result from poor germination and tillering under saline conditions.
- Plant sugarcane in furrows compared to traditional flat planting in the subtropics (Crop on the left is furrow and on the right is flat planted). This prevents the roots to come across the place of salt accumulation *i.e.*, the ridges. Further, irrigation with smaller quantities of water at higher frequency gives better yields as salts continue to remain primarily on the ridges thus protecting the rhizosphere from becoming more and more saline with the passage of time after irrigation.
- If possible, use good quality canal water between irrigations with poor quality water before monsoon rains to get a good crop.
- If organic manures @ 4–6 tonnes/acre is used, a good crop can be obtained. Use of Dhaincha as a green manure crop every alternate year would keep the land fertile. As microbial population/activity gets reduced under salt affected conditions, organic manures help to overcome these effects.
- Use of 25% more nitrogen has also been proved to be beneficial.
- Rice crop alternating with sugarcane would be successful crop rotation.
- Use of salinity resistant Co 7717 and Co 97016 varieties would give better yields. The salt affected areas in Rohtak and Sonepat districts have been successfully growing Co 7717 variety.

18.5 EFFECTS OF SODICITY ON SUGARCANE AND SUCCESSFUL CANE PRODUCTION UNDER SODIC CONDITIONS

- Excess of sodium on the exchange sites, ESP greater than 15, high pH (above 8.2), deficiency of Ca and Zn and excess of B at some places are the main reasons for low productivity under sodic conditions.
- Sodium, when accompanied by carbonates and bicarbonates, displaces calcium and magnesium from the exchange sites and the clay colloids become dispersed as a result of which the micropores start closing down leading to poor soil physical conditions. The water movement slows down leading to a deficiency of oxygen in the rhizoshere.
- The plants face a deficiency of calcium because calcium carbonate is slightly soluble in water and gets precipitated in the rhizoshere. In some sodic soils, a hard impervious pan of calcium carbonate, which starts somewhere in the upper one meter and may be of different thickness, is formed. This further reduces the hydraulic conductivity of soils.
- A level of 32 ESP and/or pH of 9.2 have been observed to lead to nearly 50% reduction in cane yields, which is not economically sustainable.
- Co 7717 is good for sodic soils as well and is followed by CoS 767. CoJ 64 is highly sensitive and hence should not be grown under sodic conditions.
- Use the required quantity of gypsum as suggested by soil test alongwith zinc sulphate @ 10 kg/acre. As calcium displaces sodium from the exchange sites, this not only leads to calcium availability but also improves the physical conditions of the soil. A decrease in pH improves availability to the plant.
- Total zinc is not less in sodic soils but rather its availability is the real problem and only 4% of it contributes towards plant growth.
- Use of FYM or any other organic manure alongwith gypsum and zinc sulphate improves the microbial activity, which is

lower in sodic soils. Production of chelating substances by the microbes facilitates zinc uptake by the plant roots.

- Fibre content of the cane increases whereas juice content decreases at high pH in sodic soils.
- Sodium increases in the juice whereas potassium decreases.
- If there is a problem of sodic waters (water with high residual sodium carbonate–RSC or sodium adsorption ratio–SAR), plant sugarcane in furrows. Use gypsum or cane mud to reduce its effects. Use gypsum based on the quality of water. Pass water over gypsum stored in the tanks before applying to the fields.
- High RSC waters slowly make the fields sodic by bringing about changes in the soil chemical properties and thereby changes appear in its physical structure.
- RSC water may bring down the sugarcane yields by 50% which depends upon the variety being grown. Use of the required quantity of gypsum will help to completely overcome the effects of these waters on cane yields.
- Plant sugarcane in furrows particularly if the field is irrigated with high RSCs/saline waters.
- If water stagnates in sodic soils upon irrigation, use poly bag technology to raise nursery and plant them when they have at least two fully expanded leaves.

Chapter 19

Control of Flowering in Sugarcane

Sugarcane flowers during October–December in the tropical India. Though largely a varietal character, it is influenced considerably by the age of the crop, photoperiod, temperature, moisture and nutrition. Flowering is undesirable in commercial plantations as it reduces both yield and quality significantly and hence has to be suppressed. The losses in yield and quality depend on the time interval between flowering and harvest and also on weather conditions. Varieties with

Fig. 19.1: Flowering in sugarcane

less or no deterioration after flowering are beneficial to both sugarcane farmers and the sugar industry (Fig. 19.1).

19.1 METHODS OF PREVENTION OF FLOWERING

19.1.1 Growing Non–Flowering/Shy Flowering Varieties

Under ideal conditions, the flowering may range from 0–100%. The safest method for eliminating flowering in commercial fields is to grow low flowering/non–flowering, non–deteriorating varieties. A number of non–flowering varieties have been identified *e.g*. Co 8021, Co 86031, Co 87044, Co 91010, and Co 94008.

19.1.2 Altering the Date of Planting

By altering the planting date, flowering can be avoided in heavy flowering areas. For example at Coimbatore, sugarcane planted after May does not flower in October–November of the same year but flowers in the same months next year. Flowering is considerably reduced in April planted crop. Certain varieties even escape flowering in April–May planting. Adsali planting or special season planting (July to September) helps in avoiding flowering and its adverse effects. Thus, if flowering varieties need to be grown, their planting dates may be altered to prevent or reduce flowering.

19.1.3 Use of Chemicals

Chemicals like maleic hydrazide, monuron, diuron, paraquat, diquat and ethephon were found to suppress flowering and have been used on a commercial scale. The beneficial effect of suppression of flowering is more in southern states where cane continues to grow through out the year. In Co 6314, cane and sugar yields increased by 15% and 6% respectively through control of flowering by ethrel spray.

19.2 CONTROL OF FLOWERING IN VARIETY Co 6304

Treatment	*Percent flowering*	*Cane yield (t/ha)*	*Sucrose % Juice*
Unsprayed (Control)	49.6	98.6	18.8
500 ppm ethrel spray	0.7	113.4	19.4
% Increase over Control	–	15	3

The chemical was sprayed @ 500 ppm during the last week of July. Flowering was suppressed to the extent of 70–95% in Co 62175 planted during March–April by spraying ethrel @ 500 ppm during the first week of July. There was about 9% improvement in cane yield.

19.3 SUPPRESSION OF FLOWERING IN VARIETY CO 62175 BY ETHREL SPRAY

Suppression of flowering in profuse flowering varieties results in higher yield, especially in regions where there is a long time interval between flowering and harvest and conditions for growth of sugarcane are favourable during this period.

Treatment	***Flowering %***	***Cane yield (t/ha)***	***Sucrose % Juice***
Control 81.9	86.8	18.78	–
Ethrel sprayed	23.9	94.9	18.40
% Increase over Control	70.0	9.3	–

Effect of ethrel spray on sugarcane (Variety: Co 62175)

Recommendation: Ethrel (2–chloroethyl phosphonic acid), an ethylene releasing compound is available in trade names such as ethephon, agrophon in the market. Its application rate is 100 ml in 100 litres of water per acre as a mist over the crop canopy using a

Fig. 19.2: Photoperiodic control under field condition

Fig. 19.3: Photoperiodic facility

Fig. 19.4: On–farm control of flowering

Fig. 19.5: Sugarcane flowering

boom sprayer between the last week of July to third week of August depending on the latitude of the location. The chemical should be sprayed in the morning hours on rain free days. Care must be taken to avoid excess dosage as it may have certain adverse effects such as sprouting of buds. The cost of the chemical and its application is around Rs. 1000 per hectare and the expected income is Rs. 5000 – Rs. 8000 due to increase in cane and sugar yields (Fig. 19.2 to Fig. 19.5).

Chapter 20

Management of Sugarcane under Drought

Frequent drought is one of the serious limiting factors in sugarcane cultivation. Formative growth stage (60–150 days) is the most critical water demand period and drought at this stage directly affects the cane yield and juice qualit (Fig. 20.1). The loss in yield is often more than 50%. A normal sugarcane crop consumes about 250 tonnes of water per tonne of cane produced. The yield of cane is directly proportional to the amount of water transpired. Therefore, maintaining adequate soil moisture throughout the growth period is important.

Fig. 20.1: Prevention of moisture loss in harvested cane

20.1 DROUGHT SYMPTOMS

The major symptom is the drying of older leaves and stunted growth of stem, resulting in a dwarf canopy with short internodes.

The leaf thickness, leaf dry weight and leaf area ratio are highly sensitive to drought.

20.2 CROP MANAGEMENT

20.2.1 Early Planting

Planting sugarcane early in the season helps the crop to grow sufficiently before moisture stress sets in so that it can tolerate drought better. In the tropical belt, November–January planting is better than later planting. In the sub–tropics, autumn (September–October) and early spring planting (February) is better. Early planted crop also gets favourable soil moisture conditions for growth.

20.2.2 Seed Rate and Spacing

Higher seed rate achieved through closer row spacing of 60–75 cm is better than widely spaced crop. Close spacing results in higher stalk number and helps in retaining more stalks till harvest. At close spacing about 15–20 percent higher yield is obtained than 90 cm spacing.

20.2.3 Soaking Setts in Lime Water

Soaking the sugarcane setts in saturated lime solution (80 kg kiln lime *i.e.*, calcium carbonate, dissolved in 400 litres of water) for one hour before planting hardens the setts and help in obtaining higher germination and better initial crop growth.

20.2.4 Trench Planting

Planting sugarcane in deep trenches (30 cm wide, 45 cm deep and refilled with loose soil + FYM + basal fertilizers up to 30 cm) helps deep root development and efficient use of nutrients and moisture.

20.2.5 Trash Mulching

Trash mulching helps in conserving soil moisture, checks weed growth and reduces soil temperature by around 2°C. Under limited irrigation, alternate furrows may be mulched and non–mulched furrows may be irrigated. Detrashing may be done leaving only 6–7 leaves per stalk and the trash may be used for mulching. The termite problem could be overcome by Lindane dusting @ 10 kg/ha.

20.2.6 Fertilizer Management

Fertilizer application should aim at developing deep root system and adequate tillering before drought period. Therefore basal application of full dose of phosphorus along with 10% of the N and K should be done followed by first top dressing at 30 days (30% of N + 30% of K) and the second top dressing before drought at 60 days (30% of N + 30% of K). The remaining 30% N and 30% K should be applied when the drought period is over in June–July.

20.2.7 Foliar Application of Urea and Potassium

Foliar application of urea and muriate of potash each at 2.5% (2.5 kg urea + 2.5 kg MOP in 100 litres of water) at 15–20 days interval during drought maintains crop turgidity, helps in the formation of proline and checks tiller mortality and thus helps in retaining adequate shoot/stalk population until moisture condition becomes favourable.

20.2.8 Protective Irrigation

During drought, available water can be given in alternate furrows alternatively. By trash mulching alternate furrows, alternate furrows (non–mulched) can be irrigated.

20.2.9 Use of Anti–Transpirants

Kaolin acts as a reflectant and reduces transpiration loss of water. It is sprayed at 3–6% weight/volume basis.

20.2.10 Tolerant/Resistant Varieties

Sugarcane is a hardy crop and can tolerate moderate moisture stress by morphological adaptations, physiological and biochemical adjustments. The inward rolling of top canopy, wax coating on the leaf surface, greater rooting depth etc are important under drought. Some of the tolerant varieties developed for drought are listed below.

Table 20.1: Drought tolerant varieties

Co 86010	Co 90009	Co 92033
Co 86249	Co 91010	Co 94005
Co 86032	Co 91012	Co 94008
Co 87023	Co 91013	Co 96008
Co 87263	Co 91018	*Cb 96012*
Co 88042	Co 91021	Co 96019
Co 89029	Co 92009	Co 96020
Co 89030	Co 92023	Co 97007
Co 89032	Co 92030	Co 97010
Co 89036	Co 92031	Co 98013

These varieties may be tested for their suitability in drought prone areas and promising ones may be popularized.

Chapter 21

Important Diseases of Sugarcane and their Management

More than 60 diseases of sugarcane, caused by fungi, bacteria, virus, and micoplasma, have been reported from different parts of India. The most important diseases, those affect the "Indian Sugar Industry" are red rot, grassy shoot, leaf scald, ratoon stunting, red strip and mosaic. It has been estimated that diseases and pests cumulatively cause a loss of nearly 15–20% in cane yield and quality.

21.1 RED ROT *(COLLETOTRICHUM FALCATUM)*

Red rot, one of the oldest diseases of sugarcane, is also known as the "Cancer" of sugarcane. The disease has been reported from all the states except central peninsular region. Large number of promising varieties had to be discontinued by farmers due to their susceptibility to this disease (Fig. 21.1).

21.1.1 Diagnostic Symptoms

Setts and settling infection

- Red or dark brown lesions, formed on nodal region specially root primordia, on leaf scar, and growth ring. Pre–emergence death of buds or post emergence death of settling generally occurs in sub tropical belt;

Fig. 21.1: Red rot disease at early crop stage

Fig. 21.2: Characteristic red rot symptoms with white spots inside the stalks

- Dark brown to straw colored continuous lesions with intense sporulation on both the surface of the midrib of unfolded leaf (spindle) and old sheath during April to July.

21.1.2 Symptoms on Standing Crop

- Mostly the third and fourth leaf turns yellow and withers away at the tip and along the margin.
- Splitted stalks often give bad smell like alcohol.
- Splitted canes display reddish discolouration with white transverse bands.
- Sometimes in rainy seasons pink masses of spores of the fungus are present on nodal region.

- Red or dark brown lesions are also formed on internodal region.
- Browning/reddening of root primordia, leaf scar and buds are also observed.

21.1.3 Foliage Symptoms

- The pathogens produce tiny 0.5 to 2 mm reddish lesions on the upper surface of lamina.
- Infection on the mid rib can be noticed from June onward. These lesions start as minute red spot on the upper surface of the midrib and develop in both directions. Lesions later become straw colored in the centre with dark reddish margins (Fig. 21.2).

21.1.4 Management

Steps like domestic quarantine, healthy seed production programme, varietal balance, regular monitoring and rouging of sugarcane crop, field hygiene, proper irrigation and the drainage in the field would be effective and helpful to minimize the spread of red rot and losses in cane and sugar production. Rotation with rice, green manure crops, linseed, onion, garlic etc. may be helpful to reduce red rot pathogen inoculum. Cultivation of resistant varieties is the most effective way to manage the red rot disease.

21.2 WILT DISEASE: (*FUSARIUM SACCHARI* AND *ACREMONIUM* SP.)

Upto 1945 the disease was mostly confined in Bihar and from there the disease spread to U.P., Punjab, Haryana, Uttar Pradesh, Uttranchal, Gujarat, Tamilnadu, Madhya Pradesh, Andhra Pradesh, and Rajasthan. Many excellent varieties had succumbed to wilt disease (Fig. 21.3).

21.2.1 Diagnostic Symptoms

- Stunting, yellowing and unthrifty appearance of few isolated clumps at the six month stage.

- The most typical symptom is the yellowing and drying of the top leaves.
- At the advance stage rapid drying and loss in weight of stalks is observed.
- In splitted canes, in early stages of the infection, the tissues of lower internodes have diffuse purple, light pink or red colour.
- There are no white cross bands and alcoholic smell as noticed in red rot disease.
- Spindle shaped cavities develope in internodal and nodal region in advance stage.

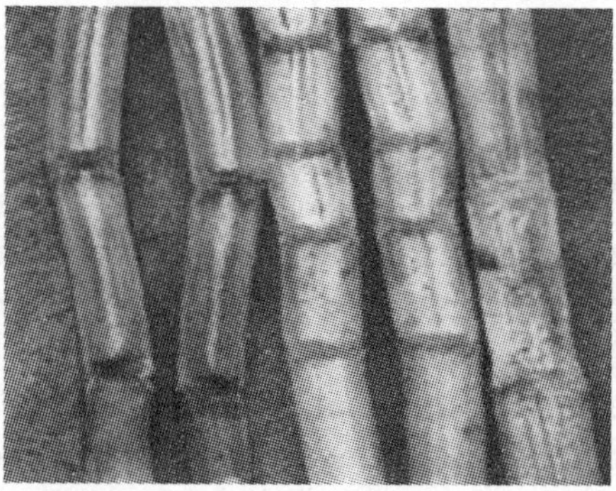

Fig. 21.3: Internal symptoms of wilt disease–affected canes

21.2.2 Management

Incidence of wilt disease would be minimized by domestic quarantine, healthy seed cane production and field hygiene. Crop rotation with rice may be helpful in reducing wilt pathogen. Sugarcane grown with onion, garlic, coriander, and mustard reduce wilt incidence. Sett treatment with boric acid 0.2% plus *Trichoderma viride* (10^6 spore/ml) for 2 to 5 minutes; application of neem cake 1.5 q/acre plus *Trichoderma viride* (mixed with neem cake) at the time of planting showed reduction in wilt and root borer incidence.

21.3 SMUT *(USTILAGO SCITAMINEA)*

Smut has been widely distributed from all the major sugarcane growing areas, but it is more prevalent in tropical region, *i.e.*,

Tamilnadu, Andhra Pradesh, Maharashtra, Karnataka, Gujarat, and Kerala (Fig. 21.4).

21.3.1 Diagnostic Symptoms

- Smut is easily diagnosed by appearance of a black whip like structure from the central core of the affected plant.
- The thin white papery membrane of the smutted whips easily ruptures expressing dense black powdery dust of spores.
- Affected plants usually have straight slender and thin stalks. These plants are taller and stand distinctly higher than the rest of the plants. Affected plants produce more tillers (Fig. 21.5).

Fig. 21.4: Smut disease affected sugarcane clump

Fig. 21.5: Smut sori (whip smut)

21.3.2 Management

Following steps would be effective to minimize losses by smut disease:

- Rouging and destruction of infected plants.
- Avoiding the ratooning of affected plant crops.
- Healthy and disease free sugarcane setts should be used for planting.
- Sugarcane seed material treated in hot water at 50°C for 2 hours eliminate smut infection in seed.
- Sett treatment with hot water with Bayleton showed effective control.
- Rotation with lucern, maize, green manuring fallow may be helpful to reduce smut infection.
- Intercropping with *Cajanus cajan* reduces secondary spread of smut.

21.4 RATOON STUNTING DISEASE

Ratoon stunting disease is caused by a bacterial pathogen. Losses are directly linked with the intensity of the disease.

21.4.1 Diagnostic Symptoms

- Setts from disease infected stalk may germinate slowly and erratically.
- Presence of stunting and unthrifty growth of plants. Diseased plant generally displays few and thin stalk with shortened internodes and yellow leaves.
- Typical symptoms are seen after splitting the cane stalk longitudinally. Vascular bundles with yellow, orange, pink red, reddish brown colour appear as dots, straight or bent forms upto 2–3 mm in length and may be just below the region of attachment of leaf sheath due to plugging of xylem vessels with coloured gummy substances.

21.4.2 Integrated Management System

- Replacement of susceptible varieties.

- Sanitation is important in preventing healthy canes free from infection.
- All the implements should be cleaned by hot water or by chemical like 5%
- Lysol, 1% Dettol or 10% Formalin.
- Heat treatment of seed cane with hot water (50°C for 2 hours) or moist hot air (54°C for 2 hours)

21.5 LEAF SCALD *(XANTHOMONAS ALBILINEANS)*

Leaf scald is a bacterial vascular disease of sugarcane.

21.5.1 Diagnostic Symptoms

Symptoms at two phases of leaf scald disease are described as under.

21.5.2 Chronic phase

- Appearance of white pencil line streaks about 1–2 mm wide on the leaf lamina parallel to veins.
- Leaf streaks often extend to the tip or margin of the leaves where they cause a withering and necrosis of the leaf tissue. Affected leaves start drying from the tip to downward and in due course the entire leaf show scalded appearance.
- Splitted canes show bright to dark red streaks of vascular bundles in the pith.
- Production of galls, multiple buds and shortening of internodes

21.5.3 Acute phase

- Sudden wilting and death of mature plants without showing any external symptoms.

21.5.4 Integrated Management System

- Growing resistant varieties in the disease prone area.
- Strict quarantine measure on the transportation of seed cane material.

- Moist hot air treatment of cane seed at 54°C for 4 hours.
- Burning of the left over plant material after harvesting.
- Ratooning of the infected plant crop should be avoided.

21.6 GRASSY SHOOT DISEASE (MYCOPLASMA LIKE ORGANISM)

It is known as new chlorotic albino and yellowing disease of sugarcane. The disease is widely prevalent in all parts of India.

21.6.1 Diagnostic Symptoms

- Stunting, profuse bushy tillering appearance and yellowing of leaves with short internodes easily recognize the disease (Fig. 21.6).
- The most common symptom on the leaves is light yellow white strip parallel to the midrib.
- Thin, weak and small canes are produced in severely infected clumps.

Fig. 21.6: Characteristic grassy shoot symptoms with chlorotic tillers in sugarcane

21.6.2 Integrated Management System

- Use of disease free and heat–treated seed setts in moist hot air (54°C for 4 hours) or hot water (50°C for 2 hours).
- Strict surveillance and rouging must be practiced.

21.7 YELLOW LEAF DISEASE OF SUGARCANE

Yellow leaf disease (YLD) is a recently identified disease of sugarcane in India. The disease is also referred to as yellow leaf syndrome (YLS).

21.7.1 Symptoms

The symptoms appear initially on third to fifth leaves from top in a maturing plant or ratoon crop. The symptoms could be very clear after 5 to 6 months of crop growth. On the leaves, the symptom appears as yellowish midrib on the lower surface. The yellowing may be confined to midrib region or the yellow discolouration may spread laterally to adjoining laminar region parallel to midrib upto a distance of 2.0 cm. Reddish colouration of midrib and laminar region is also noticed in certain varieties. In most susceptible varieties, typical yellowing of midribs and laminar region is noticed on upper surface of the leaves. Finally symptoms of necrosis of discoloured laminar region from leaf tip to bottom and subsequent drying of entire leaf is noticed (Fig. 21.7).

Intensity of the disease may vary from 0 to 85 % in the field. In ratoon crop the intensity of the disease will be higher than in plant crop. YLD incidence in sugarcane was found aggravated by poor maintenance of the crop in the field. It is observed that infestation with internode borer, flowering, drought conditions, *Striga*

Fig. 21.7: Yellow discolouration of laminar region alongwith mid rib in YLS affected sugarcane

infestation, and infection with other pathogens such as red rot, wilt, RSD, GSD etc. influence the disease development.

21.7.2 Causative Agent

The ELISA studies revealed sugarcane yellow leaf YITUS (SCYLV) as the causative agent of the disease.

21.7.3 Growth of Affected Plants

The sugarcane varieties showing mild symptoms usually record normal cane growth. In severely infected clumps cane thickness and stalk height are significantly affected. Severe infection of the disease leads to shortening of internodes. This effect culminates in bunching of leaves at the top. Usually such infection results in drying of entire clumps. Severe infection of SCYLV leads to reduced juice quality and sugar recovery. It has been proved that combined infection of SCYLV and ratoon stunting bacterium in sugarcane causes severe stunting than their infection alone.

21.7.4 Transmission

Primary transmission of the disease occurs through infected setts. Secondary transmission is achieved through aphids. Planting of infected setts shows cent percent expression of the disease in the field and build up of high virus titre in most susceptible varieties causes drying of all canes by 10–12 months.

21.7.5 Management

(i)* Seed *selection

Since primary transmission of the disease takes place through infected setts, planting of disease–free materials is required to prevent disease introduction in the field. Seed cane plots are needed to be monitored periodically for the disease at regular intervals. Seed canes should be selected from such carefully monitored fields. Alternatively, going for tissue culture seedlings derived through meristem culture ensures virus free planting materials.

21.8 MANAGING DISEASES AND STRESSES

The disease is aggravated by many biotic and abiotic factors. Biotic factors such as infestation of internode borer or *Striga* favours disease appearance and partial wilt infection induces disease expression. Additionally water logging and drought predispose the plant in expressing the disease. Since many of these biotic and abiotic factors predispose the crop in disease build up, adequate care should be taken to minimize them.

Chapter 22

Important Insects of Sugarcane and their Management

22.1 SUGARCANE WOOLLY APHID

The sugarcane woolly aphid (SWA) first appeared in Vellore in Tamil Nadu during March 2004 and has spread to all the districts bordering Karnataka and Andhra Pradesh within six months. Since SWA is new to Tamil Nadu and is fast spreading, extensive and intensive monitoring of sugarcane fields for its occurrence is imperative. For this, the involvement of sugar industry and farming community is of paramount importance.

22.1.1 Nature and Symptoms of Damage

SWA attacks all varieties and all stages of sugarcane crop. Attacked patch of' sugarcane crop appears white from a distance. Black sooty mould grows on the honeydew deposited by the aphid over the lower leaves and trash. The ground below SWA infested sugarcane patch shows white powdery coating due to the deposition of white covering from the body of SWA (Fig. 22.1).

22.1.2 Extent of Damage

Yield and quality losses range from 20 to 25%. Crops infested for more than 4 months do not sprout when ratooned. Infested cane

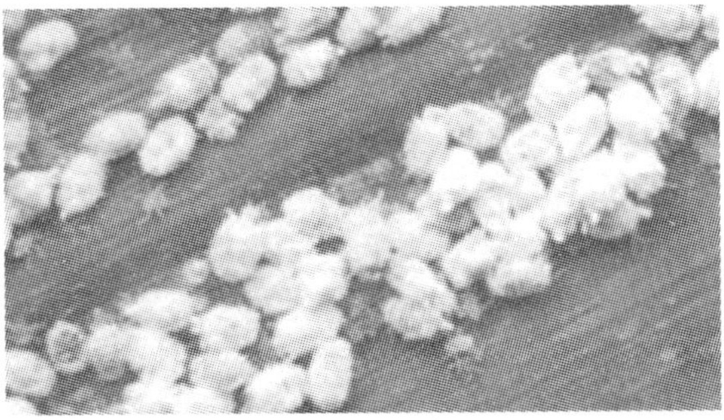

Fig. 22.1: Close–up view of sugarcane woolly aphid

tops loose fodder value either due to the presence of vast numbers of SWA or sooty mould or due to residues of insecticides applied for its control (Figs. 22.2 & 22.3).

Fig. 22.2: Woolly aphid infested field

Fig. 22.3: Sooty mould growth on lower leaves of woolly aphid infested plant

22.1.3 Life History

The aphid population consists of only females. They reproduce asexually and produce young ones called nymphs instead of laying eggs (Figs. 22.4 & 22.5). Nymphs pass through four stages in 6 to 22 days to become wingless or winged adults. First and second instar

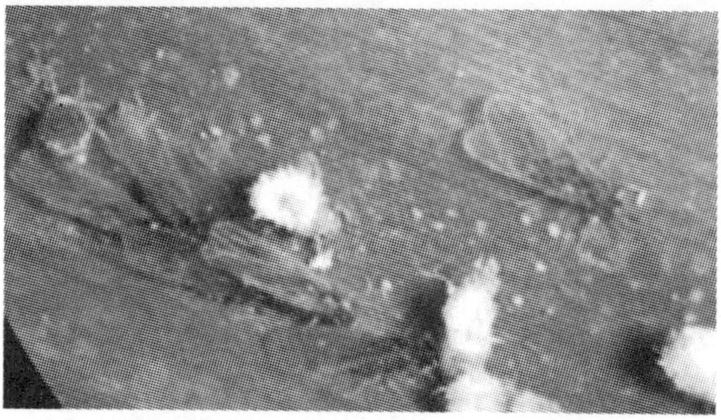

Fig. 22.4: Late females of woolly aphid

Fig. 22.5: Ballooning of woolly aphid nymphs for dispersal

nymphs show sparse woolly growth whereas winged adults do not have woolly coating on their body. Adults live for 20 days. Due to the peculiar reproductive nature, one nymph is capable of multiplying to about 220 crores of individuals in month's time.

22.1.4 Ecology

July to February is the most favorable period whereas summer (April–June) is the least favorable. Rain temporarily suppresses SWA spread. Crop that receives excess or late application of nitrogen or excess irrigation suffers more attack. Dense and shaded sugarcane crop or lodged crop harbors more population of SWA.

22.1.5 Preparative Steps to Manage SWA

Due to its invasive nature, rapid multiplication rate and fast spreading ability, intensive monitoring of sugarcane fields is the foremost step in its management. To facilitate routine monitoring of the tall stature and dense canopied sugarcane crop, the following points should be borne in mind.

1. Infested tops should never be shifted from one location to another to prevent the spread of the aphid.
2. Wide/paired row planting, periodical detrashing and propping up the cane would minimize aphid incidence, besides facilitating monitoring and spraying operations.
3. Excess or late application of nitrogenous fertilizers or excess irrigation should be avoided to minimize SWA infestation.
4. The field should invariably be monitored every week for the presence of the aphid.

22.1.6 Curative Management

Newly invaded areas
Weekly monitoring and spot application of insecticides reduce the rate of spread and establishment of aphids in newer areas.

Established areas
Harmonious use of insecticides and predators is the key to the aphid's management. This involves monitoring aphid intensity, conservation of predators and spot insecticidal applications.

Collection of infested leaves and destruction of aphid by burning or burying leaves should not be attempted as it might favour dispersal of the aphid. Before spraying infested patches, predators from all infested spots should be collected and released in one patch to serve as refugia. Leaving the refugium, the remaining patches should be sprayed with insecticides such as acephate 75 SP or monocrotophos. 36 EC or dimethoate 30 EC @ 2 g or 2 ml per litre water. Insecticides give temporary control lasting for about 15 to 20 days even if used as per recommended dose and procedure. For a permanent check, the aphid's predominant predators should be collected and conserved in refugia in the attacked sugarcane field itself.

22.1.7 Role of Farmers

1. Monitoring the field at weekly interval for the occurrence of SWA.
2. Creating refugia in the field to conserve and multiply predators.
3. Spraying insecticides in all aphid infested patches only leaving the refugia.
4. Creating awareness about the aphid among neighbouring sugarcane farmers.
5. Management of the aphid on campaign basis.

22.1.8 Role of Sugar Factories

1. Creating awareness among registered and non–registered sugarcane farmers through meetings and leaflets.
1. Ensuring the availability of fresh stock of recommended insecticides with ISI certification.
2. Ensuring the availability of adequate hand/rocker sprayers.
3. Raising the predators in shade/net houses and distributing them to farmers to create refugia.

22.2 NEMATODE

Plant–parasitic nematodes are microscopic roundworms, which feed on sugarcane roots. In India, five nematode genera, *viz., Pratylenchus, Meloidogyne, Hoplolaimus, Tylenchorynchus* and *Heicotylenchus* are found widely prevalent in sugarcane fields. Of these lesions nematode *(Pratylenchus)* is the most predominant and economically important genus followed by root–knot nematode *(Meloidogyne)*.

22.2.1 Damage

Nematode feeding disturbs the water and nutrient absorption

machinery and biochemical pathways in the root system, which in turn lead to stunted growth, chlorosis and wilting of plants. Microscopic examination of roots of plants infected by endo–parasitic nematodes such as root–lesion and root–knot nematode reveals presence of necrotic lesions and small galls, respectively on roots and nematodes inside the roots. It is estimated that nematodes cause an average annual yield loss of 15.30% worldwide in sugarcane. Among the 20 life sustaining crops of the world, maximum monetary loss due to nematodes is reported in sugarcane. In India nematodes are reported to cause 10–40% yield loss in sugarcane.

22.2.2 Diagnosis of Nematode Infection

Nematodes do not kill their hosts instantly, but debilitate them gradually and make them vulnerable to secondary pathogens and abiotic stresses resulting in disease–complexes or syndromes. Because of the silent and debilitating effect on the host without any apparent aboveground symptoms, nematode infection often goes unnoticed. Therefore, nematodes are aptly called **'hidden enemies of the farmers'**. Even in the advanced stages of infection, the aboveground symptoms are often confused with nutrient deficiencies or physiological disorders thus making diagnosis more difficult. Therefore whenever such symptoms are noticed in the field, it is appropriate to ascertain the association of nematodes by estimating soil and root population of nematodes for taking timely control measures.

22.2.3 Nematode Management in Sugarcane

Sugarcane crop requires high doses and/or multiple application of presently available chemical nematicides in order to get desired level of nematode suppression owing to its long duration. This makes the chemical control of nematodes in sugarcane economically unsound and unsustainable. This necessitates the adoption of eco–friendly integrated Nematode Management (NM) practices.

22.2.4 Eco–friendly Approaches for Nematode Management in Sugarcane

i) Green–manuring

Green–manuring with legumes like sunhemp or Daincha brings down the population of plant parasitic nematodes. Green–manure crop could be grown before planting the sugarcane and incorporated in soil during field preparation. Alternatively, it could be shown along ridges at the time of planting sugarcane and incorporated in soil 45 days after planting along with first earthing up operation. Green–manuring improves the soil organic matter, which facilitates the rapid build up of natural enemies of nematodes. Use of poor host or trap crops like sunhemp for green–manuring will enhance the effectiveness of the treatment against root–knot and lesion nematodes.

ii) Applications of press mud or FYM

Application of cured pressmud or Farm Yard Manure (FMY) @ 25 t/ha during field preparation helps in the suppression of plant parasitic nematodes. Pressmud serves as a substrate for the multiplication of nematode antagonistic microbes.

iii) Trash mulching

Mulching with cane trash @ 5t/ha and its incorporation into soil during earthing–up operation will bring down the population of nematodes. Cane trash serves as a substrate for the multiplication of nematode antagonistic microbes. Some substances released during the decomposition of trash are directly toxic to nematodes.

iv) Application of oil cakes

Application of neem or castor oil cake @ 2t/ha during field preparation brings down the nematode population. These oil cakes serve as substrates for the multiplication of nematode antagonistic microbes and contains alkaloids toxic to nematodes.

v) Intercropping and crop rotation

Intercropping with short duration legumes like soyabean and green gram and incorporation of their residues in soil after harvest reduces the nematode population. Nematode antagonistic microbes multiplied on the decomposing crop residues and toxic metabolites released during the decomposition process suppress the nematode populations. Crop rotation with poor host or non–host crops like paddy, mustard, coriander, sunflower, marigold will bring down the population of lesion and root–knot nematodes.

vi) Application of biological control agents

Application of fungal biocontrol agents like *Paecilomyces lilacinus* or *Trichoderma viride* (mass multiplied on cheaper substrates like sugarcane molasses and pressmud) @ 20 kg/ha will help in suppressing the plant parasitic nematodes, rapid decomposition of organic matter and nutrient mineralization in sugarcane field.

vii) Nematicides

In certain situations, where none of the above non–chemical practices could be implemented when field is heavily infested with nematodes, nematicides like carbofuran or phorate @ 3 kg a.i./ha could be used to quickly bring down the nematode population. Nematicides are to be applied in two splits *i.e.* half of the dose at the time of planting and the second half 90 days after planting. However, it should be followed up with application of organic amendments or green manuring in the next season to sustain the nematode control achieved using nematicides.

Farmers may adopt one or more of these practices depending on the resource availability and feasibility. Adoption of most of the above non–chemical management practices not only reduce the nematode damage but also improve soil and plant health.

22.3 SEX PHEROMONES OF SUGARCANE–INTERNODE BORER

Pheromones are the chemical languages of insects to communicate within their species. Of the many kinds of pheromones, sex pheromone is the one produced mostly by female moths to which male moths of the same species get attracted for courtship from distances as far as one kilometre.

For sugarcane ecosystem, sex pheromones are the practical pest management tools. The crop phenotype is not suited for spraying or dusting operations and the pest habits render the operation uneconomical or ineffective. Pheromones have quick action like insecticides and the crop and pest features are no barriers to their effectiveness at all. Though resistant varieties are the best management tools against pests, the complexities of sugarcane genetics and the priorities of breeding mandates almost make it impossible to develop a variety resistant to a pest with all other desirable attributes. Even if such a variety is developed, it may not suit all the agro–climatic regions wherever the target pest is prevalent and it may not be resistant to other key pests of the region. Pheromones on the other hand, operate across regions and simultaneous tackling of different pest species in one and the same trap is possible. The advantages over the bio–agents are that, pheromones once identified, can be produced on an industrial scale. As time becomes a minimum barrier for synthesis, mass production, modification, transportation and storage of pheromones, it is much easier to handle. It is effective round the year and even at low pest densities. Being species specific and safe, it is highly compatible with all the components of IPM.

22.4 INTERNODE BORER (INB): *CHILO SACCHARIPHAGUS INDICUS* KAPUR

It is a major pest of sugarcane in peninsular India. In recent years, it has modified its feeding habit. It has extended its feeding region from the top tender internodes to the spindle across the apical meristem. In this type of attack yield and quality loss is severe and the jaggery quality is also affected.

The economic threshold level (ETL) falls below 10% from 25%. The present management practices of detrashing and inundative releases of *Trichogramma chilonis* are not sufficient to contain the pest below the revised ETL.

(i) ***INB Sex Pheromones:*** Sugarcane Breeding Institute has done pioneering work on sex pheromones of various moth borers of sugarcane. INB pheromone was identified as Z–13

Octadecenyl acetate and Z–13–Octadecenyl alcohol and was found effective at 7:1 ratio in mass attraction of male moths of INB in field trials. However, due to want of indigenous know–how on mass production of the pheromone, it could not be put to field use.

(ii) ***Field Application:*** Mass trapping of male moths. In this method the male moths are attracted to a trapping system using synthetic pheromone lures of INB. Among different types of traps only water trap was found effective against sugarcane moth borers. The synthetic pheromones are impregnated at 3mg strength in rubber septa which act as the lure. Eight to ten water traps housing the synthetic pheromone septa are required for one acre. The traps are suspended at spindle height on 5th moth of the crop in a 30 metre grid. Well propped cane clumps are used as support for suspending the traps. At 9th month the trap height is reset to the spindle height as maximum efficiency was observed at spindle height. A thin film of engine waste oil is added to the water in the trap to trap and kill the moths. The septa are changed at 7th and 9th month and if needed at 11th month. The water trap is to be maintained by removing the dead moths and recharging with water and engine waste oil at regular intervals to have maximum efficacy.

22.4.1 Advantages of Pheromones in INB Management

1. Economical. The cost of 10 × 3 septa with ten SBI Wuri water traps may cost about Rs. 400/– per acre.
2. Detrashing exclusively to manage INB at 5th and 7th month can be dispensed with.

The timing of *T. chilonis* release in accordance with the moth catch will improve the per cent parasitisation of INB eggs.

Chapter 23

Farmer's Guide to Growing Organic Sugarcane

This guide for organic sugarcane production will facilitate farmers to adopt standardized organic production practices to meet the certification requirements.

Fig. 23.1: Sugarcane field

23.1 TRANSITION TO ORGANIC PRODUCTION SYSTEM

Establishment of an organic management system and building of soil fertility requires an interim period, known as the conversion period. In the case of sugarcane, a safe period of two years is recommended for transition/conversion period within which all the actions required to meet the requirements of certification is attained.

When the transition is made in a planned manner, attending to all the nutritional needs of the crop through recommended organic inputs and providing a healthy soil, organic farming will outdo conventional production system in the long run. The following are the organic production technologies for sugarcane growth in tropical zone (Fig. 23.1).

Crop Rotation

Avoid growing sugarcane in the same land. Rotation of crop's like rice, banana, turmeric, groundnut, sunnhemp and *dhaincha* with sugarcane will help in optimal use of soil resources and sustained higher cane yield.

Varieties

Recommended varieties for organic sugarcane production in Tamil Nadu are Co 8021, Co 86032, Co 86249, CoC 90063, Cog 93076, CoG 94077 and CoSi 95071.

Land Preparation

Deep ploughing once or twice with disc plough followed by shallow ploughing three or four times using a cultivator.

Spacing

Adopt a minimum row spacing of 90 cm. For fertile lands and for profusely tillering varieties, the spacing can be increased upto 150 cm. furrows must be 20–30 cm deep.

Organic Manure

Apply farm yard manure or compost or well decomposed pressmud at 80 t/ha either before last ploughing or in the furrows before planting. However the quality of organic manure could be adjusted in such a way to supply 280 kg N/ha through one or more sources like farm yard manure, compost, pressmud etc., depending upon their N content.

Planting Material

Collect setts from 6–8 months old disease free nursery crop. Two budded setts are better than three–budded setts. It is always advisable to collect the seed material from organically grown sugarcane crop.

Sett Rate and Planting

75,000 two–budded setts per hectare for 90 cm spacing. Plant two two–budded setts for every 30 cm length of furrow, in 90 cm spaced furrows.

Green Manure Intercrop

Sow green manure crops like *dhaincha* and sunnhemp, on one side of the ridges on 3rd or 4th day after planting sugarcane and raise it as an intercrop with sugarcane. Harvest and *in situ* incorporate the intercrop around 45 days after planting.

Weed Management

Hand hoeing and weeding at 30, 60 and 90 days after planting. Follow only non–chemical weed management technologies like hand weeding and mechanical weed control methods.

Biofertilizers

Apply 5 kg each *of Azospirillum* and phosphobacteria respectively on 30 and 60 days after planting of sugarcane. Mix the biofertilizers thoroughly with 500 kg/ha of farm yard manure to increase the bulkiness and apply. Give light earthing up and irrigate immediately.

Trashing

Remove the dried and senescent leaves at 5th and 7th month and apply as mulch in alternate furrows.

Irrigation Schedule

Table 1: Irrigation schedule for sugarcane

Growth stages	***Interval between irrigations(days)***
Germination (upto 35 days)	7
Tillering (36–100 days)	10
Grand growth (101–270 days)	7
Maturity (271 days to harvest)	15

The above schedule is for medium type of soils. Reduce the intervals for light soils and increase for heavy soils. When there is rain adjust the interval depending on the amount of rainfall. Ridges and furrows method is cheap and best. Convey the irrigation water from source to the field head through pipelines to reduce convenience loss.

Prevention of Lodging

At 7th month after trashing, a wet earthing up will help to reduce lodging of canes. Tying the canes with trash–twist (trash twist. propping) will also help to reduce lodging.

23.2 PLANT PROTECTION

Early shoot borer: Trash mulching, frequent irrigations and light earthing up at 35th day will result in lower incidence. Release 125 fertilized female *Strumiopsis* parasite/ha when the crop is at age of 45 to 60 days.

Internode borer: Cards pasted with 0.2 cc eggs of *Trichogramma chilonis* parasite are available in the parasite breeding laboratories. Get these cards and tie them in the field @ 25 cards/ha equally distributed in 25 places once in 15 days when the crop is 4–11 months old. Alternatively, set up pheromone traps in the field @ 25/ha spaced at 20 meters apart when the crop is 5 month old, trap and kill the male moths of internode borer. Replace the pheromone vials in the traps in 7th and 9th months.

Red rot disease: In places prone to red rot disease grow only resistant varieties such as Co 8021, Co 85019, Co 86010, Co 86032, Co *t\6249,* Co 93009 and Co 94008. In case susceptible varieties are grown, adopt the following practices: 1) Select and use disease free setts, 2) Remove and destroy affected clumps, 3) Prevent the flow of irrigation/rain water from affected fields to healthy fields, 4) Do not ratoon crop from the disease affected crop, and 5) After the harvest affected crop, grow rice crop and destroy the soil debris inoculum.

Smut disease: To manage this disease: 1) Get setts from disease free canes, 2) Remove and destroy affected clumps, 3) Do not allow more than one ratoon crop, and 4) Grow resistant varieties.

Grassy shoot disease: Treating the setts in an aerated steam therapy (AST) unit at 50°C for one hour can destroy the disease causing organism in the setts. Use the setts from 3–tier nursery raised using AST treated setts to avoid the disease.

23.3 CANE HARVEST

Harvest the canes when they are fully mature. The sucrose content in juice of a mature crop will be more than 16% and the purity of the juice will be more than 85%. In general, harvesting at the age of around 12 months is advantageous. Harvest the cane 2 to 3 cm below the ground level using a hand axe. Topping should be done at the point of break.

Cane Yield

When all the packages of practices are carried out appropriately in time, the cane yield will be around 150 t/ha. In well–drained–fertile deep soils, the cane yield can go upto 250 t/ha.

Ratoon Cane Yield

If the ratoon crop is managed well with all the appropriate package of practices, the cane yield from the ratoon crop will be equal or marginally lower (around 5%) compared to that of the previous plant/ ratoon crop.

Chapter 24

Bottling of Sugarcane Juice for Beverage

Fresh sugarcane juice obtained from mature cane is sweet and tasty. It is not only a thirst quenching drink but also nutritious due to the presence of several minerals like phosphorus, calcium, magnesium, vitamins, amino acids etc. It has, medicinal properties also particularly for the cure of jaundice. Cane juice mixed with the extracts of lemon, ginger, chillies etc adds to the taste, helps in better digestion and improved bowel movement. But due to its seasonal availability and fast deterioration by microbial decomposition, the possibility of its use as fresh juice is very much limited. In order to make the availability of cane juice at all times as a soft drink prepared under hygienic conditions, a simple method of bottling the cane juice has been developed.

24.1 CHOOSING A GOOD VARIETY

The sugarcane variety for juice should give high in juice extraction and have high sugar content. The internodes should be long without splits. The rind should be preferably light coloured and the cane should have less fibre and pith. Some of the varieties which are ideally suited for bottling of juice are CoC 671, Co 62175, Co 86249, Co 7717, Co 86032 and Co 94012.

24.2 GROWING THE CANE FOR BEVERAGE PURPOSE

It is better to grow the cane in normal soil (avoid saline/alkaline soils, since the juice taste will be affected due to excessive salts in the juice) with organic manures, as these promote better juice quality parameters. The crop must be protected from infestation of pests (scales, mealy bugs, borers) and diseases (red rot and smut) by proper plant protection measures as they affect the juice composition and taste. Avoid excessive inorganic fertilizers application. Lodging of canes in the field should be prevented. Avoid saline water for irrigation.

24.3 HARVESTING THE CANE FOR JUICE EXTRACTION

The canes should be harvested when they are at peak maturity *i.e.*, when the sucrose content is maximum and the non–sugars are at minimum. After harvesting, the top few immature internodes should be removed as they contribute to, bad taste due to high amount of salts and other undesirable organic substances. The canes should be cleaned free of roots at the nodes and adhering soil particles. It is preferable to remove the grind portion of the cane by suitable means as the pigments in the grind contribute to dark colour and unpleasant taste to the juice. The canes should be crushed in clean power operated crusher (preferably with stainless steel rollers) and juice collected after filtration through muslin cloth into a clean sterilized stainless steel vessel. One medium sized lemon and 2–3 gm of ginger can be added for every 3 kg of cane while the canes are crushed Fig. 24.1.

24.4 PROCEDURE FOR BOTTLING THE JUICE

24.4.1 Materials Required

(a) Glass bottles of 300 ml capacity
(b) Stainless steel vessels of 30–40 lit capacity
(c) Cane crusher
(d) Gas/electrical heaters
(e) Sterilizer/hot water tank
(f) Bottle corking machine

24.4.2 Juice Treatment

The filtered clear juice taken in a stainless steel vessel is heated to 60–70°C and the juice is maintained at that temperature for about

15 minutes. Due to this, impurities in the juice start coagulating. A part of the coagulated material that floats on the top of the juice is scooped out with the help of a ladle or strainer. Another part of precipitated material starts settling at the bottom of the vessel. After 15 minutes, the clear supernatant juice is decanted through a muslin cloth. While the juice is still at bearable warmth, sodium meta–bi-sulphite is added as a preservative @ 1 g per 8 litre of juice and stirred well. When the juice is still warm, it is transferred to hot water sterilized bottles and sealed with sterilized corks with a help of a corking machine. Chilling before consumption improves the taste (Fig. 24.2).

Fig. 24.1: Sugarcane juice

24.5 STORAGE PERIOD

The juice bottled as above was found to keep well at room temperature without much deterioration for at least 6–8 weeks and the taste was also not affected. Storing bottles under refrigerated condition will prolong the life for a further period of 3–4 weeks.

24.6 PRECAUTIONS TO BE TAKEN

1. The cane material should be clean and the crusher, bottles, vessels etc should be completely sterilized to prevent deterioration of juice by microbial decomposition.

2. The juice should not be heated above 80°C and should not be heated beyond 15 minutes at that temperature. Bottling of juice should be done before the juice is completely cooled to room temperature.

24.7 ECONOMICS

The initial investment for bottling 500 nos. of juice bottles daily is approximately. Rs. 5000/– to Rs. 6000/–. The cost of 300 ml quality of juice will work out to be Rs. 3–4/– approximately.

Fig. 24.2: Sugarcane juice (bottle)

Chapter 25

Computer Based Extension Modules

Modern Information Technology (IT) in the place of traditional extension approach is the need of the day in the growing context of globalization of agriculture and allied sectors. The generation and dissemination of information has to be all encompassing, demand driven and responsive to the needs of the intended clients. The ongoing IT revolution has opened up uncommon opportunities such as multimedia, internet, digital bulletin boards, electronic mail discussion groups, decision support systems (expert system), etc to developing countries in terms of providing low cost access to information. The pioneering attempts made at Sugarcane Breeding Institute in developing computer based modules are detailed below:

25.1 MULTIMEDIA ON SUGARCANE PRODUCTION

An interactive multimedia module on sugarcane production has been developed. It describes and illustrates the several steps involved in scientific sugarcane cultivation and is in the information kiosk format targeted for sugarcane growers and extension personnel involved in sugarcane production. The module envisages the use of different multimedia building blocks such as text, audio, video, graphics and animation. Macromedia Flash was used as the integrating software. The module contains detailed information from sett planting to harvest with, suitable visual clippings allowing the learner to learn at his own pace, The module has three parts *viz*., sugarcane varieties, crop production (both for plant and ration crop) and crop protection technologies (pest, disease and nematodes). The module comes in the form of Compact Disc and runs for nearly–one hour duration.

25.2 SUGAR–EX, AN EXPERT SYSTEM ON SUGARCANE PESTS

Sugar–Ex, an expert system developed by the Institute illustrates the use of decision support systems broadly in the area of sugarcane production through development of a prototype, taking into consideration the major pests limiting sugarcane yield. The expert system helps to diagnose pests for sugarcane crop, provides details of the identified pest and suggests preventive/curative measures. This prototype is a result of joint efforts by the Scientists of Extension and Entomology Sections of the Institute. The subject matter expertise on sugarcane entomology was obtained from the previous publications of the Institute, reference books and personal experiences.

The following pests have been included in the system for identification and suggesting preventive and curative measures– early shoot borer, internodes borer, top borer, stalk borer, plassey borer, gurdaspur borer, root borer, white grub, scale insect, mealy bug, white fly, pyrilla, wooly aphid, blister mite, red mite and white mite.

25.3 THE LOGIC FLOW

The brief logic flow of the expert system is as follows **Basic input:** The basic input can be given as

1. Part of the plant where damages symptoms are seen–stem/ cane, leaf, root.
2. Description of the damage symptom.
3. Life stages of the pest–egg, larva, nymph, pupa, adult.
4. Description of the life stage.

Considering these symptoms/description of the life stages, the pest is identified The user then has the option to either exit the window or further reset and proceed with diagnosis for other pests. After diagnosis, the user can directly get hyperlinked to know more about the pest, get management measures as well as get a graphic view of the damage symptoms/life stages of the pest.

All these operations are done using the "mouse", which helps a new user to get acquainted with the system faster and operate it easily. In this system, computer, graphics, digitized pictures and text are being used following a "Decision Tree" logic.

Chapter 26

Plant Tissue Culture Industry in India

India is an agricultural country and its economy is largely dependent on agriculture. Therefore, technological developments in the agriculture sector have a direct impact on country's economy. The abundance of genetic base (biodiversity), diverse agroclimatic zones and highly qualified manpower offer a potential scope for technological advances in agricultural biotechnology in India.

India has also been traditionally a leading exporter of primary, agricultural produce such as tea, coffee, rubber, cardamon, cotton, spices, flavouring, aromatic and medicinal plants etc. This status is declining due to low productivity, non–competitive prices and sometimes, poor quality of the material. Therefore, our urgent need to improve plant cultivars for quality and quantity is required using biotechnological methods.

Most of the agricultural produce in India is produced as rain fed crops. The shortage of healthy seeds/planting material and lack of disease resistant clones have often been affecting the agricultural economy. Large scale production of elite clones through micro-propagation, or somatic embryogenesis, production of disease free clones and improvement of varieties through cellular, molecular and conventional breeding methods hold the promise of improving some of the problems of Indian agriculture, horticulture and forestry.

26.1 SCOPE OF DEVELOPMENT

26.1.1 Diverse Agroclimatic Regions

India has rich resource of land spread over diverse agroclimatic zones that have the potential to produce a wide variety of agricultural and commercially high value crops. It has large biodiversity, germplasm of rare ornamentals and medicinal plants, most of them being native to this continent. Therefore, India offers many advantages over other countries in the world.

26.1.2 Forest Wealth

The native forests all over the country are well known for their richness and variety in foliage and ornamental plants. These plants can be propagated by tissue culture method and introduced in the national international markets.

26.1.3 Different Climatic Zones

In India all the climatic zones, from temperate to tropical are available. Therefore, any plant growing in any climate can be multiplied and grown in the country. On the same time, cool climatic zones are less expensive, for micropropagation work as extensive air conditioning for plant growth is not required.

26.1.4 Scientific Talent

In India, the scientific talent and the operational skills of the people is at par with the west in the world. The salaries are low in India. This makes it an ideal place for production facility and that is why, some companies have been shifted to India from USA and Europe. It is worth mentioning here that plant biotechnologies for anther and pollen culture for haploids and *in vitro* fertilization were discovered in India. Embryo culture and rescue of hybrid embryo is of quite significance in this aspect.

Propagation of plants through tissue culture has become an important and popular technique to multiply crops that are otherwise difficult to propagate conventionally by seeds and/or vegetative methods. Differentiated cells or organs are manipulated to give rise to multiple copies of the parent plant under optimum aseptic environmental conditions and appropriate stimuli. It offers several advantages over conventional propagation methods such as rapid multiplication of valuable genotypes, rapid improvement of crop plants, production of disease and virus free plants, production throughout the year is possible, germplasm conservation and easy transport and transfer (international) of material (*in vitro* material is conamination free). There are several methods developed in last 3 decades for vegetative multiplication of plant species involving axillary budding, adventitious budding, somatic embryogenesis and organogenesis from callus cultures. In fact, the genetic stability of the plant copies (population) produced is likely to vary according to the procedure chosen. However, these methods have certain limitations.

26.1.5 Problems Associated with Agricultural Biotechnology

Development and use of any technology depends on infra–structural facility and simultaneous development of other related technologies. Agricultural biotechnology requires infrastructure like electricity, air–conditioned rapid transport system, skilled labour and related biotechnology like innovative and novel packaging material, plastic ware and glass appliances etc. Followings are some major hurdles -

1. Industry is a capital intensive (requires elaborate establishment money).
2. Seasonal activity with peak demand; delivery of plant material at one time requires large facility which may be under utilized at other times.
3. High production cost applicable to selected crop plants with established protocols for multiplication.
4. Management of personnel and organization of work schedule in chain oriented manner.

5. Contamination problem adds a big cost factor.
6. Somaclonal variation at large multiplication rate.

In context of India, on export scene, lack of variety of packing material at affordable price, high air freights, unsuitable air connections, lengthy phytosanitory, custom and export documentation formalities make this non–traditional, highly perishable export commodity less attractive as compared to potential it has. However, recently the government of India has set up suitable cargo handling cold storage facility at New Delhi and by Bangalore for export of biotech products.

26.1.6 Commercialization

Commercialization of agricultural biotechnology began in India in real sense in 1987 when various private sector companies adopted this technology as a means of increasing the productivity of their primary produces or as a part of their diversification programme. This includes A.V. Thomas group of companies, Indo–American Hybrid Seeds (IAHS), Hindustan Lever, Tata Tea, Unicorn Biotech, Nath Seeds, RP Goenka Enterprises. Tobacco Company, Maharashtra Hybrid Seeds Company, Hindustan Agrigenetics Ltd. and some others.

The first professionally managed commercial plant tissue culture unit in the country was set up in 1987–88 by A.V. Thomas, at Cochin in export processing zone, and remained one of the largest in Asia and only one of its kind in India (producing 190 million plants in 1996), until IAHS began production in its Bangalore based unit. IAHS has 48 Laminar flow work stations, hundred growth rooms (each can accommodate 400,000 culture at one time) and a green house of 30,000 sq. ft. (2790 sq m) with most advanced equipments such as automatic shade system, fogging, blackout and irrigation systems, and movable benches far hardening of tissue culture plants.

Subsequently, advancement in commercialization of plant tissue culture and acceptance of plantlets raised through tissue culture have led to continued exponential growth within the industry in terms of number of new units as well as number of plants produced by these units (Table 1). Most of these units are set up with foreign

Table 26.1: Plant micropropagation industry in india.

State	*No. of Industries*	*Annual Production (capacity in million)*
Maharashtra	25	More than 31
Karnataka	9	30.1
West Bengal	9	5
Andhra Pradesh	6	16.25
Tamil Nadu	4	26.35
Kerala	4	25.35
Haryana	3	13.25
Gujarat	3	11.25
U.P.	3	25.0
Punjab	1	1.7
H.P.	1	5
Orissa	1	Not known
Rajasthan	1	Not known

know how. Most of our commercial units prefer to seek the entire technology package on a turn–key basis. The major foreign collaborators (mostly with buy back arrangements) are M/S Cultiss, Holland; M/S Green Teek, Holland; M/S Microplants Ltd., UK; Centre de Research Agronomique, Belgium; Planex, Australia; and Agrobio, France etc.

Indo–American Hybrid Seeds (IAHS), A.V. Thomas and Co., Harrisons Malayam Ltd. Godrej Plant Biotech Ltd. and SPIC have established ultramodern micropropagation facilities. IAHS and A.V. Thomas have exported millions of tissue cultured plants to Europe and released several lakhs banana and cardamom plants to farmers in domestic market. A.V. Thomas has developed a cardamom variety with yield of 250 kg/hectare as compared to normal 70 kg/hectare and can be cropped in two years instead of usual three years. A.V. Thomas has also developed high glucovanillin producing and early flowering vanilla plants through tissue culture. Cost effective micropropagation system has also been developed in tea and coffee.

IAHS has won several contracts from the Netherlands, UK and Denmark to develop new plant varieties by somatic hybridization, selection of somaclonal variants, endosperm culture, *in vitro* pollination and embryo rescue method in combination with conventional breeding. IAHS is providing service for screening plants for 40 viruses, 2 viroids and several critical, fungal and mycoplasmal diseases for 19 ornamental and 26 horticultural and agricultural crops. This company had exported cut flowers worth Rs. 14 million to Holland Denmark and UK in 1991.

Beena Nursery (Pvt.) Ltd. Trivandrum offers high quality and disease free plantlets of orchids suitable for the cut flower industry. Hindustan Lever Ltd. Mumbai is involved in clonal multiplication of high yielding cultivars of plantation crops such as cardamom through meristem culture, virus free micro tubers of potato. It has also designed a bioreactor to scale–up the micro tuber production. Godrej Plant Biotech Ltd. (earlier known as Unicon Biotech), multiply trees and plants by apical and axillary meristems and somatic embryogenesis. Micropropagation protocols for 2 cultivars of strawberry suitable for cultivation in tropical plains and conventional hilly regions have been developed. The company sold over 2 lakhs plants to farmers in Maharashtra, Gujarat and Andhra Pradesh during 1992–94.

Micropropagated banana, *Chrysanthemum,* rose, carnation and foliage ornamental plants are not only being produced at large scale but also being exported. Dhampur Sugar Mills, Dhampur and EID Parry, Chennai are involved in production of virus–free sugarcane through tissue culture.

SPIC Science Foundation, Chennai has introduced sex–typed female papaya plants in commercial tissue culture market for the first time in India. SPIC Agro Biotech Centre is working on coffee, banana, potato, sunflower, rose, gerbera, orchids and tree species (*Callistemon*) jack fruit, mango and date palm).

National Agricultural and Scientific Research Foundation, Calcutta is engaged in large scale production of *Musa, Spathiphyllum, Anthurium, Elettaria* and *orchids.* It has already developed protocols for tea, jojoba, lilies, gladiolus, gerbera, *Ficus* and *Philodendron.* In all there are 75 units of small, medium and large capacity, capable of producing 190 million plants per annum. The major 4 units are producing 15–25 million plants per annum. The others are capable of producing plants from 1–10 million per annum.

26.2 PLANTS UNDER MICROPROPAGATION

Indian units are currently producing fruit crops, forest trees, ornamental (foliage and flowering plants), vegetable crops and plantation crops (Table 2). The major emphasis is on the production of ornamentals.

Table 26.2: Plant species currently micropropagated at commercial level in India.

Fruit crops	Banana, Citrus, Grapes, Papaya, Pomegranate, Strawberry, Pineapple, Cherry.
Vegetative crops	Potato, *Asparagus, Brasica napus. B. juncea,* tomato, cauliflolwer, cabbage.
Plantation crops	Cardamom, Sugarcane, Vanilla, Coffee, Turmeric, Tea, Black pepper.
Ornamentals flowering plants	*Alstroemeria, Anthurium, callality, Carnation, Chrysanthemum,* Gerbera, Impatiens, Rose, Gladiolus, Iris, orchids, Lily, *Dendrobium.*
Ornamentals foliage	*Aglaonema, Calathea, Cordyline, Diffenbachia, Ficus, Philodendron, Spathiphyllum, Syngonium.*
Forest trees	Callistemon, *Eucalyptus,* Teak, Kadamb, Jack fruit, *Simarouba* Salvadora and Bamboo.

26.3 GOVERNMENTAL EFFORTS

The Indian Government has identified micropropagation of plants through tissue culture as an industrial activity under Industrial (Development and Research) act, of 1951 which was made effective in 1991. The units engaged in floriculture can avail duty free imports under export oriented units (EOU)/export processing zone (EPZ) scheme even if they sell 50% in domestic market and export the remaining 50%. The import duty on live trees, plants, bulbs, roots, cut flowers as well as ornamental and foliage plants has been reduced from 55% to 10%. Import duty on seed development machinery for soil preparation, greenhouses, pre–cooling unit and refrigerated transport units has been reduced to 25%. A subsidy on air–freight to a maximum of 25% has been allowed for export of cut flowers and tissue culture plants.

Government of India has formed National Board of Biotechnology in 1982 to promote research and development work in the area of Biotechnology in the country. Since then several well defined programmes of National priorities began in the country through National Institutes and Universities. Later on, Department of Biotechnology (DBT) was set up in February 1986 for better functioning under the Ministry of Science and Technology. This department is planning, promoting and coordinating biotechnological programmes in the country.

DBT has set up task forces in various biotechnological areas. The research projects funded by DBT are in various research themes like micropropagation for fuel, fodder and biomass, medicinal and endangered plant propagation, protoplasts culture and genetic engineering, and crop improvement. Plant species have been identified for Research and Development programmes. Two pilot plants for micropropagation have been set up at National Chemical Laboratory, Poona and Tata Energy Research Institute, Delhi.

Government of India has liberalized rules for import of basic materials, provides tax exemptions and subsidies on air freight of plants raised through tissue culture, cold storage facilities have been created at Delhi, Mumbai and Bangalore for storage of consignment awaiting shipment. Tissue culture consignments are given quarantine clearance within 6 hours. Due to these reasons, Indian export of floriculture products increased from Rs. 738 lakhs in 1990–91 to Rs. 1139 lakhs in 1991–92.

CHAPTER 27

Plant Tissue Culture: Impact on Industry

27.1 IMPACT ON INDUSTRY

In last 20 years, plant tissue culture and related biotechnology has attained great dimensions. These technologies have already been put to industrial use in certain areas and others are being explored. All the research on plant material is directed towards economic problems. The global biotechnology business is estimated to be around 150 billion US dollars of which 50–60% is in agribusiness. Annual demand of tissue culture raised products constitutes about 10% of the total, *i.e.*, 15 billion US dollars with annual growth rate of about 15%. The advantages of plant biotechnology over conventional methods of plant improvement (breeding) are numerous. The production of botanical materials would be possible independent of environmental factors (climate, pests and seasons), it would be possible to produce them on any place of the earth thus removing the impact of political crisis and foreign competition. The quality of raw material would reach uniformity never observed with field–grown material and any new discovered drug plant can be put in culture and biomass can be produced in very short period as compared to conventional propagation and field trials. Following are the major areas of research towards proven economic implications.

27.2 MICROPROPAGATION

The micropropagation industry came to be recognized as global profit making agriculture based system in 1980s and expanded exponentially during 1985 to 1991. During this period global annual production capacity increased from 130 to 513 million plants. Of the 100 commercial units in North America, five major units produce 15–20 million plants per year, 8–10 units have annual production capacity of about 8–10 million plants, and rest 85% small units are producing less than 1 million plants per year. Of 248 commercial units in Western Europe with production capacity of 212 million plants, only 37 produce more than 1 million plants per year. In the Netherlands, there are about 67 commercial units producing 62 million plants annually. Five of them have annual production capacity of more than 5 million plants. In Germany of total 24 units, four have production capacity of 1–5 million plants. Other companies in these countries are small producer of micropropagated plants in varying numbers.

In India 21 commercial units in 1994 with targeted production capacity of 80 million plants and 14 units under construction with annual capacity of 60 million plants were visualized. In all 75 units are working presently. The four major producer companies are producing plants in the tune of 15 to 25 million. Three units with capacity of 10 million, 12 units with production capacity of 5–6 million and 4 units with 1–3 million plants are established. In a total 190 million plants per annum can be produced by these units.

Micropropagation of plant through plant tissue culture technique has already attained the dimension of a large plant based industry in several countries. This is the first plant industry based on biotechnology and is almost 30 years old. Commercial application of plant tissue culture started in USA with micropropagation of orchids in 1970s. There has been tremendous increase globally from 1985 to 1990 in the number of production units as well as the number of plants produced. With an estimated global market of 15 billion US dollars per annum for tissue cultured products, the demand far exceeds production. In developing countries, labour is available with lower wages; so plants are being produced at much cheaper rates. Indian micropropagation has expanded exponentially from 5 million plants annual capacity in 1988 to 190 million in 1996. The facilities created are at par with the west in leading countries like the

Netherlands and USA. This is a labour intensive industry and automation of various steps is in progress. Because of labour intensive work, more and more industries from the developed countries are shifting to the third world countries where labour is still cheap. The advantages of this technology over conventional propagation are–

1. Small area requirement
2. Large number of plants produced in short duration
3. Cloning of selected material is possible
4. Easy to handle and transport and does not require phytosanitary regulation.

Table 27.1: Micropropagation of plants through tissue culture

Flower crops	***Ornamental foliage***
Anthurium andre	*Asparagus*
A. scherzerianum	*Caladium*
Antirhinuin majus	*Coleus blumei*
Begonia	*Dieffenbachia*
Chrysanthemum morifolium	*Dracaena* spp.
Dianthus caryophyllus	*Ficus* spp.
Eschscholtzia californica	*Monstera deliciosa*
Euphorbia pulcherrima	*Scindapsus aureus*
Gerbera jamesonii	*Woodwardia fibriata*
Gladiolus	**Succulents and other ornamentals**
Hemerocallis sp.	*Ascelepias*
Iris hybrids	*Bougainvillea*
Lilium spp	*Clerodendrum*
Macleaya cordata	Cycas
Narcissus sp.	*Haworthia*
Pelargonium hortorum	*Jatropha*
Petunia hybrida	*Mammilaria*
Rose sp.	*Passiflora*
Tulipa gesneriana	*Tylophora zamia*
Vegetables and condiments	**Fruits and Nuts**
Allium cepa	*Achrus zapota*
A. sativum	*Actidinia chinensis*
Apium graveolens	*Aegle marmelos*
Brassica campestris	*Ananus comosus*
B. oleracea	*Carica papaya*
B. napus	*Citrus* spp.

Table 27.1: (*Contd...*)

Table 27.1: *(Contd...)*

Capsicum annuum	*Coffea arabica*
Carum carvi	*Musa cavendishii*
Cichorium endivia	*Phoenix dactylifera*
C. intybus	*Punica granatum*
Cucurbita pepo	*Prunus cerasus*
Cucumis melo	*P. hybrids*
Curcuma longa	*Rubus* sp.
Daucus carota	*Vitis vinifera*
Foeniculum vulgare	*Ziziphus* spp.
Lactuca sativa	**Forest trees**
Lycopersicon esculentum	*Albizzia lebbeck*
Petroselinum hortense	*Abies balsamea*
Phaseolus vulgaris	*Anogeissus spp.*
Piper nigrum	*Azadirachta indica*
P. longa	*Biota onentalis*
Pisum sativum	*Cupressus anzonica*
Solanum melongena	*Dalbergia sissoo*
S. nigrum	*D. latifolia*
S. tuberosum	*Eucalyptus* spp.
Trigonella foenumgraecum	*Feronia limoia*
Zingiber officinale	*Picea glauca*
	P. tremuloides
	Pinus spp.
	Populus tremula
	Prosopis spp.
	Santalum album
	Sterculia urenes
	Tecomella undulata
	Tectona grandis

27.2.1 Horticulture

Plant tissue culture has been successfully employed to propagate horticultural plants and attained a status of medium sized successful industry. Banana, strawberry, pomegranate, citrus, papaya, apple and related species have been successfully micropropagated in India and abroad. Millions of plants of strawberry and banana are produced in India by some tissue culture based industry like Jain Irrigation; Indo–American hybrid seeds Co. Bangalore; Maharashtra Hybrid Seed Co. Jalna; Unicorn; Godrej; Hindustan Lever etc.

27.2.2 Floriculture

Micropropagation is the most widely used application of plant biotechnology in floriculture. Cut flowers produced from plants obtained through micropropagation are a flourishing business in many countries. In India, there are many companies engazed in this business. The principal materials are Carnation, rose, gladiolus, jasmine, heliconias and ornamental foliage plants etc. In most cases, superior germplasm is provided by foreign collaborator to the Indian firm and final product is purchased by foreigner company (buy back arrangement). Orchids is another example where plant biotechnology has provided means to produce large number of plants from immature seeds, otherwise difficult to produce plants. As a result of this, orchid flowers are available in the market. Orchid was the first material multiplied through plant tissue culture. Orchids fetch very high cost in the market due to their beauty and their long vase life.

27.2.3 Vegetables

Improved varieties of potato, cauliflower, chilly, asparagus and sugar beet are being produced through plant biotechnological methods. Once the variety is developed by any method it is multiplied through biotechnological methods. Varietal improvement and propagation have become a multimillion dollar business in the world.

27.2.4 Sugar Industry

Plant biotechnology has contributed significantly in the development of virus–free sugarcane plants, somaclonal variants in sugarcane, somatic and genetic improvement in sugar beet towards direct industrial application of the technology in sugar industry. Sugarcane Planters Association, Hawaii contributes significantly in the research funding for sugarcane improvement (Fig. 27.1).

27.2.5 Forestry

In USA and Europe, gymnosperms (because of temperate forests; picea, *Abies, Pinus, Cedrus*) are being propagated at commercial level to replenish the forests. In India, *Eucalyptus,* Poplar and bamboos are being propagated at commercial level for the use in pulp manufacturing in paper industry and match stick production in match factories. The pilot–scale plant of Tata Energy Research Institute at New Delhi is producing other trees of forestry like *Anageissus,* and National Chemical Laboratory, Pune is producing *Tectona grandis* and sandal wood.

27.2.6 Fibre Industry

The work on cotton and jute has provided new varieties for their respective industries. The fundamental work on cotton ovule has provided insight about the development process in cotton.

27.2.7 Spices

Cardomom and clove are principal crops being micropropogated through tissue cultures. Besides these ginger and the turmeric are improved and micro-propagated. *In vitro* rhizome formation and development of disease–free plants have been achieved to improve ginger cultivars.

27.3 VARIETAL IMPROVEMENT

The improvement of a cultivar variety is a continuous process and plant biotechnological methods contribute significantly in this endeavour. Varieties can be improved through exploitation of somaclonal variation, induced variation through use of chemical mutagens of gamma radiation, resume of hybrid embryos, culture of immature embryos and through selection for various resistances. Improvement of varieties has a direct bearing on production availability, superior characteristics and consequently on market price *e.g., Brassica* species, cotton, rice, potato, sugarcane, etc.

27.4 GENETIC TRANSFORMATION

Plant biotechnology offers various methods to improve plant materials through genetic changes. These methods have revolutionized the microbe based drug industry (antibiotics, vaccines, insulin etc.). In plants therefore several examples are there where new technology has changed the crop yield in quality or quantity.

The most prominent one is the transfer of *Bt* gene from *Bacillus thuringiensis* in tobacco, tomato, cotton and other species.

Plants have become resistant to the insect after transfer of *Bt* gene and varieties fetch high market price. The seed industry is growing in all the countries, developed and developing. In this attempt to protect the right of inventor, terminator gene has been incorporated in crop plants (seeds containing terminator gene will fail to germinate in next generation) and it is likely to be a multi-million dollar industry.

Change in flower colour in carnation from white to blue by modifying the gene for chalcone synthesis, transfer of cholera vaccine gene in potato and banana, change in keeping quality of tomato are examples where biotechnology has changed the agriculture industry.

27.5 HAPLOIDS AND HOMOZYGOUS LINES

Discovery of haploid production by Guha and Maheshwari (1964) has opened new avenues of producing homozygous lines for further breeding work using conventional methods. As stated above, seed industry is flourishing throughout the world. This industry is benefited by haploid plant production through biotech methods leading to new improved varieties.

27.6 VIRUS–FREE PLANTS

The pioneer work of Morel showed that apical meristems are free from viruses. Development of virus free plants through meristem culture has improved the sugarcane and potato varieties and other species (Table 2). Disease free plants and improvement have direct effect on market price and profit to related companies.

27.7 SECONDARY METABOLITES

Secondary products are the major sources of pharmaceuticals, food additives, fragrances and pesticides. It is estimated that the market potential for herbal drugs in the western world alone could range

Table 27.2: Virus free plants through meristem culture

Crop	***Pathogen eliminated***
Allium sativum	Garlic mosaic virus
Asparagus officinalis	Viruses A, B, and C
Brassica oleracea	Cauliflower and Turnip mosaic virus
Citrus sp.	Exocortis viroid, stubborn spiroplasma
Dahlia	Dahlia Mosaic virus
Dianthus barbatus	Carnation latent
Manihot esculenta	Cassava mosaic virus
Nicotiana rustica	TMV, Cucumber mosaic virus
N. tabaccum	TMV
Petunia	TMV
Saccharum officinarum	Sugarcane mosaic virus
Solanum tuberosum	Potato virus A, G, M, S, X, T etc.

from US$ 4.9 billion to 47 billion by the year 2000 if the AIDS epidemic continue unchecked. A similar situation exists for plant based food additives, fragrances and biopesticides. The natural resources for plant based chemicals are limited but the demand for natural products is greater.

27.7.1 Direct Production

Biotechnology plays important role in producing secondary metabolites of interest by alternative means. Presently, about half a dozen chemicals are produced commercially through biotechnological means. Shikonin derivative (red dye used in cosmetics) from *Lithospennum erythrorhizon in* Japan, ginsenosides from ginseng roots, methyl digoxin by biotransformation of methyl digiotoxin by cell cultures of *Digitalis lanata* are the products produced at industrial level using plant cell biotechnology. There are several compounds which are likely to be produced at the industrial level with more innovations leading to cheaper technology (Table 27.3). Therefore, this field is a promising one for the commercial production of important secondary metabolites. High cost and great demand of the pharmaceuticals (Table 27.4) forced the scientists to look for alternative means of production. Plant biotechnology holds promise

Table 27.3: Commercially produced secondary metabolites through plant tissue culture.

Product	***Plant species***	***Biological activity***
Ajmalicine	*Catharathus roseus*	Hypotensive
Artemisin	*Artemisia annua*	Antimalarial
Camptothecin	*Camptotheca acuminate, Nothopodytes foetida*	Anticancer
Castenospermine	*Castanospermum australe*	Anti AIDS
Forskolin	*Coleus forskohlii*	Cardiotonic
Digitoxin, digoxin	*Digitalis purpurea, D. lanata*	Cardio–vascular
Shikonin	*Lithospermum erythrorhizon*	Red dye
Taxol	*Taxus brevifolia, T. baccata*	Anti–cancer (ovarian)

Table 27.4: Pharmaceutically important drugs for plant tissue culture industry

Product	***Plant***
Ajmalicine	*Catharanthus roseus*
Vinblastine	*C. roseus*
Vincristine	*C. roseus*
Ajmaline	*Rauwolfia serpentina*
Vincamine	*Vinca minor*
Ellipticine	*Ochrosia elliptica*
Camptothecine	*Camptotheca acuminata*
Emetin~	*Cephalis ipecacuanha*
Berberine	*Coptis japonica*
Sanguinarine	*Papaver somniferum*
Colchicine	*Colchicum autunmale*
Aconitine	*Aconitum* sp.
Diosgenin	*Dioscorea deltoidea*
Ginsenosides	*Panax ginseng*
Podophyllotoxin	*Podophyllum peltatum*

for the production of such compounds. Though production in the cultures is usually low, by manipulating various factors, it is possible to increase the yields to a certain extent.

27.7.2 Plant Cell Cultures as a Source of Novel Compounds

Sometimes cultured cells have been found to accumulate compounds which were not found as constituents of the parent plant. The basis of obtaining such novel compounds in cultures is uncertain. Inadequate analysis of plant material, expression of new genes, blocking of some genes (pathway), mutation etc. are considered as reasons for record of new compounds (Table 27.5).

Table 27.5: Examples of compounds detected in cell culture which were previously unknown as constituents of the corresponding plant species

Plant species in culture	***Compound***
Andrographis paniculala	Paniculid–A
Gardenia jasminoides	Tarennosid
Ruta graveolens	Rutacultin
Picralima nitida	Pericine
Voacanga africana	Voafrine–A and B
Ochrosia elliptica	Epchrosine

27.7.3 Enzyme Potential and Biotransformations

Plant cell cultures of a species producing secondary compound or unrelated species may be used for the production of a compound through biotransformation studies. Following examples provide the information about use of cell cultures.

(a) Bioconversion of p–methyl–digitoxin into tJ–methy–digitoxin by cell cultures of *Digitalis lanata.* This process is used at the industrial level in Germany.

(b) Bioconversion of ellipticine into 5–formyl ellipticine by cell of *Choisya ternata* (an unrelated species) was achieved.

(c) The Allelix company is producing 3, 4–anhydrovinblastine from bioconversion of catharanthine and vindoline by cell free extracts of *Catharanthus roseus.* Cell cultures were unable to produce dimeric indole alkaloide *de novo.* A cell free extract (or cell free system) is an enzyme preparation from cells not containing living cells in the preparation.

All the new biotechnological developments, *viz.*, varietal improvement on any trait, production of new compounds or production of cloned materials, are important for commercial exploitation of material. All the new materials fetch very high cost in the market and provide financial benefits to the inventor. That is why more and more, private funding is available to this field and many private companies have come–up in seed production and other biotech industries.

Fig. 27.1: Sugarcane processing in sugar industry

Chapter 28

Scenario of Sugarcane in India and Abroad

Sugarcane (*Saccharum* sp. Complex) being the most valuable commercial crop of India, not only sustains sugar and distillery industries but also holds a key position in the national economy by earning foreign exchange. The contemporary sugar industry plays many significant roles in relation to food, energy and economic security. Globally approximately 150 MT of sugar is produced annually and the developing countries contribute over 70 per cent of the total sugar production. Sugar is the third most important source of dietary energy supply and currently contributes more than 8 per cent of total caloric intake after cereals and oils. As far its economic importance in food security, sugar is a major source of export earning and nearly 30 per cent of global sugar production enters the world market valued over US$ 12 billion estimated in 1988 and US$ 24 billion estimated in 2011. The developing countries earn around US$ 6.5 billion from their sugar exports nearly 5 percent of the value of their total agricultural exports. The over all impact of sugar export earnings as a means to finance food imports is more pronounced for developing countries. Besides the current diversion of sugar crops for bioethanol production is set to change the socio–economic fabric of the developing countries and therefore, the contribution of the sugar crop to national GDP may be very significant in the imminent future. India ranks second among the sugarcane growing countries of the world in both area and production. Globally it is cultivated over an area of 20.1 million ha, with an annual production of 1318.1

million tonnes and productivity of 65.5 tonnes per ha. In India sugarcane is cultivated over an area of 4.36 million ha, with an annual production of 281.8 million tonnes and productivity 64.6 tonnes per ha. In Karnataka, sugarcane is grown over an area of 3.85 lakh hectares with an annual production of 324.79 lakh tonnes of cane and productivity of 84.4 tonnes per hectare. The sugar industry in India is second largest agro based industry next only to cotton with more than 450 sugar factories (Anonymous, 2005).

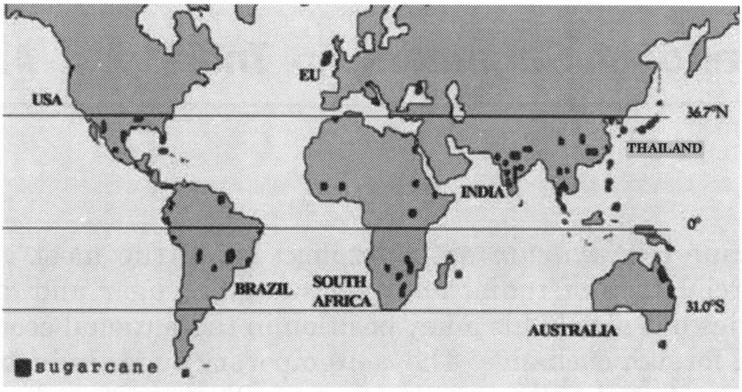

Fig. 28.1: Highest sugarcane production countries

GLOBAL DISTRIBUTION OF SUGARCANE

In India, about 50 million farmers, their dependents and a large number of agricultural laborers are involved in sugarcane cultivation, harvesting and ancillary activities constituting 7.5 per cent of the rural population (Anonymous, 2005). Although sugar is the most important product of sugarcane, various byproducts serve as industrial raw material, bagasse is used as fuel in the sugar industry itself as a fuel and paper, pulp and board are also made out of it. Molasses is an energy rich byproduct of sugar manufacture. It is used in the manufacture of alcohol, citric acid and animal feeds. The green tops of the plants obtained during harvest and thinning operations are valuable green fodder for cattle (Fig. 28.1).

There are two sugarcane belts in the country with distinct agro–climatic conditions. The tropical belt comprises of Maharashtra, Tamil Nadu, Karnataka, Andhra Pradesh, Gujarat and Kerala. The sub–tropical belt, where a large part of the cane area lies, comprises of Uttar Pradesh, Punjab, Haryana, Bihar and other north and north–eastern states. The yield levels in tropical zone are quite high (100–120 t/ha) compared to sub–tropical zone (40–50 t/ha). About 50 per cent of the cane was processed into white sugar and the remaining was used for producing jaggery, khandasari sugar and for feed purposes (Anonymous, 2005). Cane yield is markedly influenced by many factors like soil fertility, climate, variety, moisture stress, cultivation practices, weeds, insect pests and diseases. The estimated

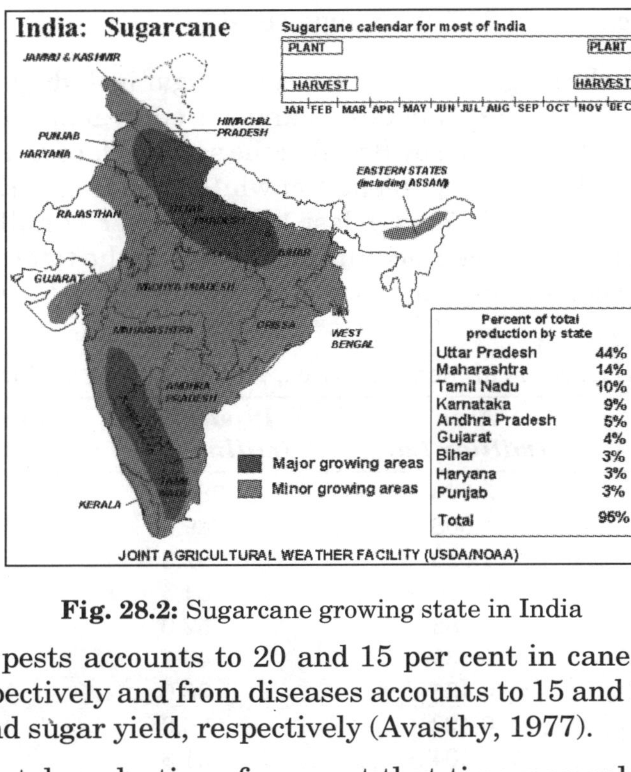

Fig. 28.2: Sugarcane growing state in India

loss from pests accounts to 20 and 15 per cent in cane and sugar yield, respectively and from diseases accounts to 15 and 10 per cent in cane and sugar yield, respectively (Avasthy, 1977).

The total production of sugar at that time was only about 1.5 lakh tonnes, whereas the consumption was about 12 lakh tonnes. To meet the domestic demand of sugar, India had to import sugar mainly from Java (Indonesia).

Sugarcane growing countries of the world are lying between the latitude 36.7° north and 31.0° south of the equator extending from tropical to subtropical zones. This map depicts the distribution of sugarcane in the world (Fig. 28.1). Worldwide sugarcane occupies an area of 20.42 million ha with a total production of 1333 million metric tons. Sugarcane area and productivity differ widely from country to country (Table 28.1). Brazil has the highest area (5.343 million ha), while Australia has the highest productivity (85.1 tons/ha). Out of 121 sugarcane–producing countries, fifteen countries (Brazil, India, China, Thailand, Pakistan, Mexico, Cuba, Columbia, Australia, USA, Philippines, South Africa, Argentina, Myanmar, Bangladesh) 86% of area and 87.1% of production (Table 28.2). Out of the total white crystal sugar production, approximately 70% comes from sugarcane and 30% from sugar beet.

India is the single largest producer of sugar including traditional cane sugar, sweeteners, khandsari and gur equivalent to 26 million tonnes raw value followed by Brazil in the second place at 18.5 million tonnes (Fig. 28.2). Even in respect of white crystal sugar, India has ranked No.1 position in 7 out of last 10 years. Over 11 million tons of refined sugar is produced, accounting for 60% of the total sugar cane cultivated. The total Indian export of the sugarcane in the year 2000–

Table 28.1: Sugarcane in the world: Area, production and productivity

Country	*Area (million ha)*	*Production (million tons)*	*Productivity (tons/ha)*
Brazil	5.343	386.2	72.3
India	4.608	289.6	62.8
China	1.328	92.3	65.5
Thailand	0.970	64.4	66.4
Pakistan	1.086	52.0	47.9
Mexico	0.639	45.1	70.6
Colombia	0.435	36.6	84.1
Australia	0.423	36.0	85.1
USA	0.404	31.3	77.5
Philippines	0.385	25.8	67.1
Indonesia	0.350	25.6	73.1
Cuba	0.654	22.9	35.0
South Africa	0.325	20.6	63.4
Argentina	0.295	19.2	65.2
Myanmar	0.165	7.5	45.4
Bangladesh	0.166	6.8	41.2
World	20.42	1333.2	65.2

Table 28.2: Sugarcane area, production, and utilization (Area in million hectares, Yield in metric tons per hectare, and others in million metric tons)

Sugar cane	*AREA/1 Million /ha*	*YIELD/1 Tons /ha*	*PRODN/1 Million tons*	*SUGAR/2 Million tons*	*KHAN-DSARI/3 Million tons*	*GUR/3 Million tons*	*SEED/3 Million tons*
1985/86	2.86	59.99	171.68	68.98	10.48	71.62	20.60
1990/91	3.69	65.39	241.05	122.32	13.18	76.63	28.93
1995/96	4.15	68.02	282.09	174.76	10.00	67.27	30.06
2000/01	4.32	68.49	295.60	176.65	11.00	72.48	35.47
2001/02	4.41	67.38	297.21	180.32	10.50	70.73	35.67
2002/03	4.52	63.58	287.38	194.33	9.50	49.07	34.49
2003/04	3.94	59.39	233.86	132.51	10.00	61.35	30.00
2004/05	3.66	64.74	237.09	124.77	9.50	74.37	28.45
2005/06	4.20	66.93	281.17	188.67	8.50	49.00	35.00
2006/07	5.15	69.03	355.52	279.30	7.50	32.73	36.00
2007/08	5.06	68.81	348.19	249.91	7.00	59.28	32.00
2008/09	4.40	64.78	285.03	144.98	6.50	100.55	33.00
2009/10	4.20	66.14	277.80	185.55	6.50	52.42	33.34
2010/11	4.94	69.25	342.38	240.00	7.50	53.80	41.08
2011/12	5.09	68.30	347.87	260.00	7.00	39.12	41.74
2012/13/3	5.25	69.52	365.00	270.00	7.00	44.20	43.80

Note: Figures for 2011/12 and 2012/13 are FAS estimates.

Source: /1: Directorate of Economic and Statistics, Ministry of Agriculture
/2: Indian Sugar Mills Association except 2011/12 and 2012/13
/3: FAS/New Delhi estimate

2001 was 811027.5 MT. Punjab is one of the major players in the production of sugarcane in India (Fig. 28.3).

Sugar is expected to play a pivotal role in the economic prosperity of developing countries, which are also significant producers of sugar crops. It is therefore imperative that the production and production efficiency of sugar crops should be sustained through new technological interventions. Besides conventional products like Gur and Khandsari, it provides raw materials of economic importance such as molasses, bagasse, wax

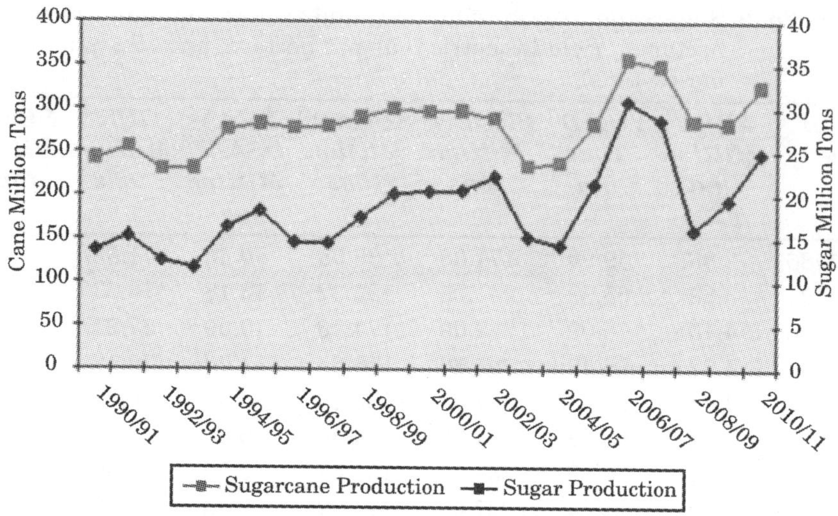

Fig. 28.3: India's sugarcane and sugar production

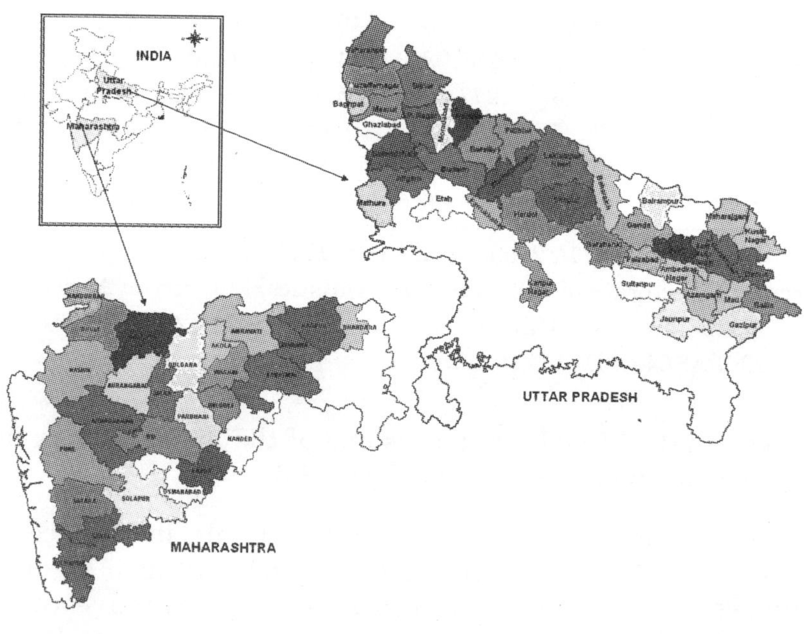

Fig. 28.4: Two States of highest sugarcane production in India

etc. to several other industries. Sugar industry is the second largest agro based industry in the country, next only to cotton and textile

Table 28.3: Sugarcane varieties over decades

	Sub–Tropical zone
Twenties	Co 205, Co 210, Co 213, Co 214, Co 224, Co 281, Co 290
Thirties	Co 205, Co 213, Co 223, Co 244, Co 281, Co 285, Co 290, Co 312, Co 313
Forties	Co 213, Co 312, Co 313, Co 331, Co 356, Co 453
Fifties	Co 312, Co 313, Co 453, Co 951, CoS 245, CoS 510
Sixties	Co 312, Co 975, Co 1107, Co 1148, BO 17, CoS 510
Seventies	Co 312, Co 1148, Co 1158, BO 17, CoS 510
Eighties	Co 1148, Co 1158, Co 7717, BO 91, BO 99, CoJ 64, CoS 687, CoS 767, Co 7314
Nineties	Co 1148, BO 91, BO 99, CoJ 64, CoJ 81, BO 128, CoLK 8001, CoPant 84211, CoS 767, CoS 802, CoS 7918, CoS 8436, CoSe 92423
	Tropical zone
Twenties	Co 213
Thirties	Co 213, Co 243, Co 281, Co 290, Co 313, PoJ 2878
Forties	Co 213, Co 41, HM 320, POJ 2878
Fifties	Co 419, Co 449, Co 527, HM 320
Sixties	Co 419, Co 527, Co 658, Co 740, Co 853, Co 975, Co 997
Seventies	Co 419, Co 527, Co 658, Co 740, Co 975, Co 997, Co 853, Co 62175, Co 6304, Co 6806, Co 6415
Eighties	Co 419, Co 740, Co 975, Co 62175, Co 6304, Co 6907, Co 7219, CoC 671, Co 8021, Co 7704, CoA 7602
Nineties	Co 740, Co 62175, Co 6304, Co 7219, CoC 671, Co 7704, Co 7527, Co 7508, Co 7504, Co 8011, Co 8014, Co 8021, Co 8208, Co 8362, Co 8371, Co 8338, Co 85004, Co 86032, Co 85019, Co 86249, CoT 8201, CoC 85061, CoC 90063, CoC 92061, CoSi 86071, MC 707 (Co 97009), CoG 93076, 85 R 186, CoSi 95071

industry. The laundry harvested about 940 million tonnes of sugarcane from about 3.7 million hectares of land during 1990–91, with an average productivity of 65.3 t/ha, which was quite lower than some leading sugarcane growing countries *viz.*, Peru (112.9 t/ha) and Egypt (100.3 t/ha). To date India is the largest producer of cane and sugar in the world but the cane and sugar production per unit area is still lower than several other cane producing countries. In India and other countries sugarcane varieties recommended for commercial cultivation in different states is given in Tables 28.4, 28.5 & 28.6.

Uttar Pradesh occupies first position in the country with regard to area as well as production of sugarcane (Fig. 28.4). However, its contribution regarding productivity at the national level has not been upto the desired extent. It appears that the state has almost reached a plateau in the yields of sugar and sugarcane. Thus sugarcane production offers a continuing challenge to increase both cane yield and sugar recovery through the development of improved clones with higher yield potential along with resistance for abiotic and biotic stresses.

Table 28.4: List of sugarcane varieties recommended for commercial cultivation in different states

S.No.	*State*	*Early ripening*	*Mid season and late ripening*
1.	Uttar Pradesh	CoS 8436, CoS 88230, CoS 95255, CoS 96268, CoSe 3234, UP 05125, CoJ 64, CoS 08272, CoS 03251, CoSe 01235, CoSe 98231	CoS 767, CoS 8432, CoS 97264, CoS 96275, CoS 97261, CoS 99259, CoS 07250, CoSe 95422, CoSe 01434, UP 0097, CoSe 01424, CoPant 84212, CoSe 96436
2.	Bihar	BO 90, BO 99, BO 102, BO 120	BO 104, CoS 767, BO 109, BO 89, BO 91, BO 106, BO 108, Co 1148
3.	West Bengal	BO 90	Co 1148, Co 7224
4.	Orissa	Co 7508, Co 7704, Co 62175, Co 740	Co 7219, CoJ 8201, Co 975, Co 706, Co 8402, Co 62175
5.	Assam	Co 1108	Co 8315, CoJor 1, CoJor 2
6.	Punjab	CoS 8436, Co 89003	CoJ 67, CoJ 83, CoJ 84
7.	Haryana	Co 89003, Co 7717, CoJ 64	Co 1148, CoS 767
8.	Rajasthan	Co 29, Co 997, Co 527, Co 6617	Co 1253, Co 419, Co 1007, CoJ 111, Co 449, Co 527
9.	Madhya Pradesh	Co 527, Co 775, Co 1305, Co 7314, CoC 671	Co 678, Co 1307, Co 6304, Co 419, Co 617, Co 1305, Co 62175, CoLK 8001, Co 7807, Co 8304, Co 7318, CoS 707, Co 1148
10.	Gujarat	Co 775, Co 975, CoC 671, Co 8338	Co 7527, Co 6304, Co 8021, Co 62175, Co 8011, Co 740

Table 28.4: (*Contd...*)

Table 28.4: *(Contd...)*

S.No.	*State*	*Early ripening*	*Mid season and late ripening*
11.	Maharashtra	Co 419, Co 775, Co 7219, CoC 671	Co 740, Co 7219, CoM 7125, Co 7527, Co 86032
12.	Karnataka	Co 6415, Co 7704, CoC 671, Co 85002	Co 62175, Co 740, Co 8014, Co 8021, Co 8011, Co 8371, Co 7804, Co 86032
13.	Andhra Pradesh	Co 6907, Co 7805, CoT 8201, Co 8013, Co 8014, CoC 671	Co 62175, CoA 7602, Co 7219, 85 R 186
14.	Tamil Nadu	CoC 671, CoC 8001, Coc 85061, Co 7704, Co 8208, CoC 92061, CoC 90063	Co 6304, CoSi 776, CoSi 86071, Co 8021, Co 85019, Co 86032, Co 86010, Co 86249, CoSi 95071, CoSi 96071, CoSi 98071, CoG 93076

Table 28.5: List of sugarcane varieties recommended for commercial cultivation in other countries

S.No.	*Country*	*Commercial varieties*
1.	Australia	Co 740
2.	Bangladesh	Co 785, Co 6311, Co 8341, Co 421, Co 1148, Co 604, Co 475, Co 642, Co 527, Co 530, Co 635, Co 1158, CoK 31, Co 775, Co 6806, Co 640, Co 631, Co 419
3.	Brazil	Co 6415, Co 331, Co 419, Co 421, Co 740, Co 775
4.	Burma	Co 419
5.	China	296
6.	Central Romana	Co 331, Co 853
7.	Cuba	Co 421, Co 312, Co 281
8.	Egypt	Co 413, Co 6304, Co 1148
9.	Florida	Co 281
10.	Indonesia	Co 997
11.	Iraq	Co 995, Co 1148, Co 6304
12.	Louisiana	Co 281, Co 290
13.	Malawi	Co 6217, Co 622, Co 419, Co 775, Co 100
14.	Mauritius	Co 6415, Co 6907, Co 7508, CoA 7701, CoA 8401, CoC 671, Co 8013, Co 92061, Co 8014, Co 85 A 261, Co 975, Co 7602, CoT 8201, CoR 8001, Co 7805, Co 7219, Co 7706, Co 6304

Table 28.5: *(Contd...)*

Table 28.5: (*Contd...*)

S.No.	*Country*	*Commercial varieties*
15.	Morocco	290
16.	Mozambique	Co 421
17.	Mexico	Co 213, Co 290
18.	Nigeria	Co 6304, Co 6415, Co 740, Co 957, Co 1001, Co 421, Co 475, Co 396, Co 765, Co 775, Co 434, Co 453, Co 961, Co 310, Co 440, Co 301, Co 285, Co 331, Co 356, Co 617, Co 349, Co 404, CoLK 30
19.	Pakistan	Co 312, Co 313, Co 1148, Co 1158, Co 740, CoL 29, CoL 44, Cp: 54, Co 421, Co 547, Co 622, Co 997, Co 658, CoJ 245, CoS 321
20.	Philippines	Co 440, Co 527
21.	Queensland	Co 475, Co 270, Co 440, Co 331
22.	Spain	Co 290, Co 413, Co 421
23.	Sri Lanka	Co 290, Co 312, Co 419, Co 421, Co 449, Co 453, Co 527, Co 617, Co 622, Co 655, Co 658, Co 617, Co 731, Co 740, Co 775, Co 785, Co 789, Co 853, Co 875, Co 876, Co 925, Co 976, Co 991, Co 997, Co 1001, Co 1111, Co 1148, Co 1163, Co 1287, Co 6304, Co 6415, Co 6403, Co 6803, Co 8232, Co 62175, CoJ 64
24.	Sudan	Co 218, Co 453, Co 527, Co 997, Co 1001, Co 6806, Co 62175, Co 945, CoS 109, Co 6502, Co 1148, Co 1158, Co 1177, Co 1007, Co 1223, Co 1230, Co 1287, CoS 510, Co 6304, Co 975, Co 785, Co 621101
25.	Thailand	Co 331, Co 419, Co 421
26.	Uganda	Co 421, Co 945, Co 6806
27.	Venezuela	Co 421
28.	Vietnam	Co 86032

Fig. 28.5: Sugarcane variety CoJ 64 Early Ripening

Fig. 28.6: Sugarcane variety CoS 99259 Mid Late Ripening

Chapter 29

Important Contributors of Plant Tissue Culture

G. HABERLANDT

According to the cell theory proposed by Schleiden and Schwarm (1839), cell is considered as structural and functional unit of living organisms. In plants, totipotency of excised plant tissues and cells was established subsequently. It was *Gottieb* Haberlandt who conceived the idea of culturing isolated plant cells in the nutrient solutions. Based on cell theory, he assumed that there was no limitation of divisibility, therefore, he started with isolated mesophyll cells using Knop's nutrient solution, sucrose, aspargine and peptone. Haberlandt (1902) in his famous paper described the cultivation of mesophyll cells of *Lamium purpureum* and *Eichhornia crassipes*, epidermal cells of *Ornithogalum* and hair cells of *Pulmonaria*. The cells survived 3–4 weeks but without cell division, there was increase in cell size. He conducted several experiments and carefully recorded his observations and interpreted the results. Using pieces of mature potato tubers he observed that cell division occurred in small, thin tissue almost without exception when the explant contained a vascular strand.

If Haberlandt had used in place of highly difficult tissues, some other materials like willow or carrot, which are known to proliferate without growth substances, he would have obtained the first tissue culture. The other reason for the failure was the incomplete information about nutrition of plant tissues and complete unawareness about plant growth regulators.

P. NOBECOURT

P. Nobecourt, a French plant pathologist, announced simultaneously in 1939 with White and Gautheret; the possibility of cultivating plant tissues for unlimited period. It was possible with the use of recently discovered Indole Acetic Acid by F.W. Went (1932). This success opened new avenues of cultivating plant cells for different studies. This success was announced a few months before the beginning of second world war. For six years, American and French scientists worked without knowledge of their mutual results.

P.R. GAUTHERET

Like many others, Pierre Roger Gautheret was working earlier at the University of Sorbonne, Paris, France. Gauther*et al*so tried to cultivate isolated cells and root tips without getting tissue cultures. Following these failure, he turned towards the tissue participating in wound healing. He first used piece of cambium cut from trees, especially *Acer pseudoplatanus, Sambucus nigra* and *Salix capraea.* Preliminary attempts with liquid medium failed completely. Later explants were placed on the surface of medium solidified with agar. He did not expect any good results and placed those cultures in a cupboard. He was surprised to find very healthy callus on the explant after two months. He observed living turgescent dividing cells. Over the next 5 years the recognition of the importance of B vitamins in yeast extract and the auxin (1M) allowed significant advances to be made. It was because of these factors that Gautheret could achieve the proliferation and division of cambial tissue of *Salix capraea* and *Populus alba* for several months. In 1939, Gautheret reported the propagation of carrot as the first plant tissue, culture of unlimited growth. IAA stimulated the growth of undifferentiated tissue on cut surfaces of sterile explants. This tissue, termed as callus, was similar in appearance to wound tissue and subsequently it was found that callus could be subcultured indefinitely. His results were immediately verified in Italy by Grelli. Finally, the possibility of cultivating plant tissues for unlimited periods was announced almost simultaneously by White (1939), Nobecourt (1939) and Gautheret (1939). In spite of difficulties encountered during the war, Gautheret was able to receive some collaborators in his laboratory and one of them was George Morel. Gautheret was working on fleshy organs, especially Chicory, Jerusalem artichoke and *Brassica napus* and Morel established strains of lianas, herbs and trees.

During the study of normal and tumour culture, the development of auxin non–requiring tissues were observed by French and American scientists. Gautheret discovered the reasons for this difference (Kulescha and Gautheret, 1948), the normal tissues maintained *in vitro* produces an insignificant amount of this growth substance. Later on, hyper–auxinity has been established in tumour tissues. The same was observed in case of habituated tissues not requiring exogenous supply of auxin. He proposed the term 'habituation' for such cultures.

In 1942, Gautheret observed development of buds in tissue culture of *Ulmus* and plantlets in Chicory tissue cultures, establishing the totipotency of the cells. The various investigations carried out by Gautheret during 30s and 40s were mainly concerned about nutrition and cell differentiation of cultivated cells. His other major contributions are establishment of hormones autonomous cultures (habituated cultures) from hormone requiring cultures and study of tumour cells. He also worked on root culture of *Helianthus tuberosus* and demonstrated that light, temperature and other factors affect the root growth. His important contributions are cited in following two monumental works.

1. Gautheret, R.J. (1955). The nutrition of plant tissue cultures: *Annual Review Plant Physiol.*, 6: 33–484.
2. Gautheret, R.J. (1983). Plant tissue culture: A History, *Bot Mag. Tokyo.* 96: 393–410.

GEORGE MOREL

G. Morel was among the first to culture monocotyledonous tissues. He developed the method of meristem culture for the elimination of viruses and the micropropagation of orchids and discovered the two unique of crown gall tissues. These opines have become marker for analysis of transformed cultures induced by *Agrobacterium tumefaciens* and *A. rhizogenes.* Morel, in 1950, obtained the indefinite growth of monocotyledonous tissues such as *Gladiolus, Iris* and lily on the medium enriched with natural extract (coconut milk, yeast extract). He also cultivated Royal Fern on this medium.

Moral and Martin (1952) developed meristem culture technique and recovered *Dahlia* shoots, free from viruses, by meristem tip culture. In 1955, they recovered virus free potato. This attained wide application of plant tissue culture to raise virus free plants in agriculture. The first industrial application was the work of George Morel who successfully multiplied tropical orchids through division of proto corms, little differentiated structures developing naturally on orchid embryos. In 1960, Moral observed the emergence of such bodies after carrying out a cymbidium shoot–tip culture. He also noted that protocorms could be sectioned into quarters and subcultured, each section regenerating a new proto corm, within a few weeks which, in turn, could be divided. The so obtained protocorms subsequently evolved into young plantlets. This remarkable property found almost immediate commercial use since orchid producers had to cope with slow multiplications rate through cluster division. Around 1970, first commercial multiplication laboratory was established in USA for orchids. In 1971, he studied *Agrobacterium tumefaciens,* crown inducing bacteria, of the dicots and isolated octopine and nopaline from *A. tumefaciens* infected plants. These chemicals are produced only when Ti–plasmid of the bacteria integrates with the genome of the host plant cells. They are also known as opines genes from the plasmid. Moral joined the laboratory of Professor R. Gautheret in France during the World War II. He established cultures of lianas, herbs and trees. Gautheret described him as able and enthusiastic worker.

PHILIP R. WHITE

P.R. White, an American scientist who worked at Rockfeller Institute, New Jersey, USA, was one of the pioneer tissue culturist of early period. In 1934 he reported for the first time successful continuous cultures of tomato root tips and obtained indefinite growth of roots. Initially he used salts of Knop's solution, sucrose and yeast extract in his medium. Later, yeast extract was replaced by three vitamins, *viz*, pyridoxine, thiamine and nicotinic acid. This synthetic medium has been proved to be one of the basic medium for a variety of tissue cultures. White maintained some of his root cultures till before his death in 1968.

With the advice of Stanley, a famous scientist who isolated tobacco mosaic virus (TMV), White started work on the dual culture. He established cultures of tomato roots infected with TMV. Cultures of infected roots gave rise to healthy roots by root tip cultures. The success of White's experiment opened the field of root cultures, which have been explored by many workers to solve morphological, physiological and pathological problems.

In 1939, White had presented his work in tobacco tissues at a meeting of the Botanical Society of America. He reported for the first time, long term callus cultures which were established from tumour tissues of *Nicotiana glaucax N longsdorfil.* He was successful due to the incorporation of auxin in the medium. Similar results were also observed by Gautheret and Nobecourt in the same year by conducting independent research on carrot. White has contributed in formulating root culture medium, which is known by his name as White's medium (1943). This medium consists of essential elements, glycine, nicotinic acid, thiamine, pyridoxine, sucrose etc. The work of White can be categorized into three aspects, continuous growth of excised tomato root tips in liquid medium, *in vitro* cultivation of viruses on excised roots and growth of tumour tissues. On March 25, 1968, he died after suffering from acute hepatitis. His book on tissue culture was the only book on the subject.

– White P.R. (1963). *The cultivation of animal and plant cells.* Ronald Press, New York.

F.C. STEWARD

F.C. Steward was one of the pioneers of plant tissue culture and contributed a lot by way of understanding the requirements of plant tissue cultures and developing techniques for the different applications. Working at Cornell University, Ithaca, USA; he had developed a School of Plant Physiologists involved in plant tissue culture work. Caplin and Steward (1948) used coconut water for the first time in plant tissue culture and obtained vigorous proliferation of carrot explants. It was also evident from this finding that the stimulating substance of coconut milk was not an auxin. Steward and co–workers were the pioneer in establishing cell cultures of single cells in liquid medium (Steward and Shantz, 1955) and developed various types of vessels for this purpose including the rotating nipple flasks.

The another landmark comes in the form of somatic embryogenesis in carrot cell suspension cultures and production of complete plants (Steward and Ammirato, 1969). Steward had taken cell suspension of callus tissues derived from wild Carrot embryos and plated them out on a coconut milk medium solidified with agar. From this plated cell population, thousands of embryo developed each being derived from one or a few cells of the callus tissue. This has not only opened new avenues of micropropagation of plant species at an enormous rapid rate but also completely established the totipotency of the cell. His volumes on 'Plant physiology, a treatise' were the most widely used reference books on plant physiology and tissue culture.

J. REINERT

J. Reinert, a German Scientist, was a pioneer in the field of plant culture and contemporary with Steward, Cocking and Street. In 1959, he was culturing undifferentiated parenchymatous cells of carrot root in complex medium containing sugar, salt and coconut milk. On this medium the callus had been sub–cultured repeatedly, without evident morphological change. He transferred the tissue to a synthetic medium containing an elaborate mixture of known substances, including amino acids, vitamins and hormones. The tissue became more granular and showed evidence of differentiated regions. This culture, on transfer to a medium lacking high levels of auxin, produced young plants. Histological analysis showed bipolar embryo formation in the system, which produced plants on proper medium. This had proved the development of a complete plant from mature undifferentiated parenchyma and in other words 'the totipotency' of cells.

Besides the embryogenesis in carrot, Reinert worked several years on developing technology for wide ranging applications of plant tissue culture, which includes cryopreservation of cells, regeneration of plantlets, anther culture, process of androgenesis, development of haploids, and protoplasts culture. The book 'Plant cell, tissue and organ culture' by Reinert and Bajaj (1977) Springer Verlag, Berlin Heidelberg, was the first compilation of state of art literature on plant tissue culture.

INDRA K. VASIL AND VIMLA VASIL

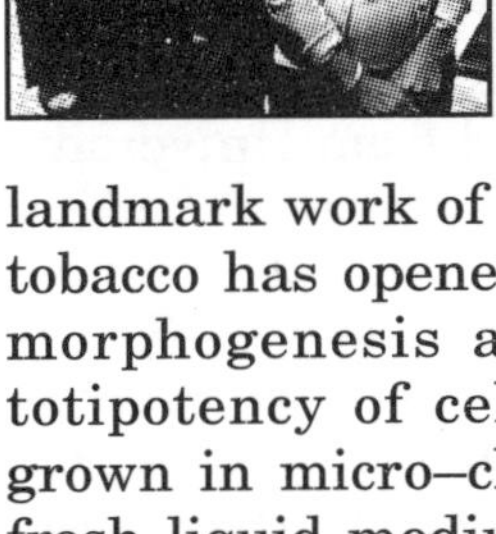

Prof. Indra K. Vasil and Dr. Vimla Vasil are the product of Delhi School of Morphology headed by the then Professor P. Maheshwari. They joined the laboratory of Professor A.C. Hildebrand, at Madison, Wisconsin, as post–doctoral fellow. Vasil, V. and Hildebrandt were the first to prove the early cell theory of Schwarm by regenerating a tobacco plant from a single cell derived from a suspension of cultured cells. The landmark work of Vimla Vasil on culture of single isolated cells of tobacco has opened new applications of cell cultures in genetics, morphogenesis and vegetative propagation and proved the totipotency of cell. Single cell of hybrid tobacco callus were grown in micro–chambers in the absence of any other cells, in fresh liquid medium containing coconut milk. These produced a colony of 50–75 cells in 10–25 days. Upon transfer from the micro–chambers to agar medium, cell clumps produced callus in 3 months, which ultimately produced complete plantlets (Vasil V. and Hildebrandt, 1965).

During his stay at Wisconsin, I.K. Vasil studied cell differentiation and morphogenetic behaviour in *Petroselinum hortense* and *Cichorium endivia.* He was successful in producing C. *endivia,* plantlets from freely suspended cells grown *in vitro.* I.K. Vasil has contributed significantly on the cell culture, protoplast cultures and somatic embryogenesis of cereals (monocotyledons), particularly *Pennisetum americanum, Panicum maximum, Triticum aestivum,* and *Zea mays.* He has edited a series of books on plant tissue culture and somatic cell genetics of higher plants published by Academic Press. Monocotyledons, particularly the cereals were considered unsuitable material to regenerate via somatic embryogenesis. I.K. Vasil and co–workers demonstrated that the young leaves, inflorescences or embryos are useful in initiating embryogenic cultures. The comprehensive work showed that selection and maintenance of embryogenic cultures at an early stage is important for maintaining long term embryogenic potential. Establishment of embryogenic and non–embryogenic lines in monocotyledons and concept about these cultures were given by I.K. Vasil. Embryogenic cultures are compact, dry, amorphous and white in colour as compared to non–embryogenic cultures which are watery, dirty white to light brown in colour and soft in nature. It is necessary to isolate and maintain separately embryogenic cultures.

During long career as University Professor, Prof. Vasil worked as editor to several prestigious scientific magazines like 'Nature Biotechnology', 'Plant Science' and 'Plant Cell Reports'. He has also edited books and contributed in these volumes, in which 'Cell culture and somatic cell genetics of plants, published by Academic Press, are widely used volumes on plant tissue and cell cultures. About 400 scientific articles were written by him. He retired from University service in 1999.

In recent years, he worked on identification, isolation and transfer of glutenin and insect resistant gene in wheat, improvement of cereals, and development of herbicide resistant wheat cultivars through use of micro projectile particle gun technology. In cereals, wheat, pearl millet, maize, barley and sugarcane are principal crops he worked on. Some of his recent publications are as follows–

1. Vasil, I.K. (1998). Biotechnology and food security for the 21st century: a real perspective. *Nature Biotechnology*. 16: 399–400.

2. Vasil, I.K. and Anderson, O.D. (1997). Genetic engineering of wheat glutein, *Trends Plant Sci.*, 2: 292–97.
3. Altpeter, F.I., Vasil, V., Srivastava, V. and Vasil, I.K. (1996). Integration and expression of the high molecular weight glutenin subunit Iax1 gene into wheat. *Nature Biotechnology*. 14: 1155–59.

RIKER AND HILDEBRANDT

Professor Albert C. Hildebrandt was a professor at the University of Wisconsin, Madison, U.S.A. As a graduate student and through ensuing years as a staff member, Hildebrandt worked closely with Professor A.J. Riker, who had a keen interest in the effect of the crown gall bacterium, *Agrobacterium tumefaciens,* on the host plants and its possible relation to cancer in animals. Professor Riker was a plant pathologist and used plant tissue culture to understand the basic biology of infection of plant pathogens (Hildebrandt and Riker, 1958). As an approach to the study of crown gall disease, Hildebrandt began the study of the growth *in vitro* of excised plant tissue, both healthy and diseased. Prof. Hildebrandt became an internationally recognized authority on tissue culture. He was very active in the international tissue culture association. He contributed chapters' to several books and manuals concerning tissue culture techniques and their uses.

The tissue culture work began with the study of the isolation and growth of tissues in culture, including the effects of nutrition and environment. Isolations were made from healthy stems, galls, apical meristems and anthers. The effects of bacteria and viruses on, the callus tissues were widely studied. Tissues were induced to differentiate back to entire plants through the use of plant growth substances. This was done with a number of plant species – tobacco, sunflower, potato, geranium, gladiolus, African violets, poplar and cassava. Hildebrandt, with his students, then developed methods for isolating single cells from tissue cultures (micro–chamber, nurse tissue culture etc.), growing them in liquid culture and again inducing them to differentiate back to the original plant (Sievert and Hildebrandt; 1965). A great deal of still and cine–photomicroscopy was done on the growth of isolated cells and their relation to disease organisms. This work on regeneration of plants from single cells (Vasil and Hildebrandt, 1965) become the basis for the eventual development of the technique for the plant transformation in the biotechnology industry. Prof. Hildebrandt also proposed for the first time the possibility of fusion of somatic cells, the somatic hybridization. Hildebrandt formulated a plant tissue culture medium known as Schenk and Hildebrandt (1971) medium. The isolation of single cells from apical meristems and anthers was used as a means to produce virus–free plants. Hildebrandt did this for several plants, including geranium, gladiolus and poplar. Virus–free stock was propagated and made available to commercial growers. Related to this work was the study of several virus diseases of gladiolus and geranium. Hildebrandt was resource person for information on diseases of floriculture crops for growers of the state. Hildebrandt was born in Pennsylvania, on April 10, 1916. He obtained his Ph.D. degree in plant pathology in 1945 and joined the faculty in 1949 as assistant professor and rose to the rank of full professor in 1960. He retired in 1978. Emeritus Prof. Hildebrandt, age 84, died on January 4, 2001.

Some important citations of the group are as follows:

1. Hildebrandt, A.C. and Riker, A.J. (1958). Viruses and single cell clones in plant tissue culture. *Feb. Proc.*, 17: 986–993.
2. Tamaki, T., Hildebrandt, A.C., Burris, R.H. and Riker, A.J. (1961). Oxidation of reduced diphosphopyridine nucleotide by mitochondria from normal and crown–gall tissue cultures of tomato. *Plant Physiology*, 36: 347–351.
3. Hildebrandt, A.C. (1977). Single cell culture, protoplasts and plant viruses. *In*: Plant Cell, Tissue, and Organ Culture, J. Reinert and YPS Bajaj, Springer–Verlag, Heidelberg.

FALKE SKOOG

Professor Folke Skoog was one of the leading plant physiologists of the world. Skoog was already renowned for his pioneering work with auxin, a plant growth hormone, when he joined the University of

Wisconsin, Madison, U.S.A. While at Wisconsin, he discovered a major new class of plant hormones, the cytokinins, which stimulate the division of plant cells, and regulate plant growth and development. During the cultivation of tobacco pith cells, Skoog and co–workers observed that the incorporation of purine base–adenine in the culture medium enhances callus growth. Similarly, DNA extracts from yeast and Herring sperm whale also favoured callus initiation and callus growth of tobacco pith cells. This leads to the isolation of a compound from autoclaved Herring sperm whale DNA called kinetin (6–furfuryl aminopurine). Though Kinetin enhances cell division in pith cells, it does not present in this form in plant cells. Therefore, they concluded that this product was formed during autoclaving of whale DNA. It is produced due to dehydration and migration of a deoxyribose moiety from the 9–position of adenine to the N6–position (Miller *et al,* 1956). His work has had a profound impact on agricultural and horticultural practices around the world.

Skoog's discovery of cytokinins triggered an international flood of publications that continues till date. His work showed that a number of cytokinins occur naturally, and that at least one of these compounds occurs in every organism tested, from bacteria to humans. With colleague Nelson J. Leonard at the University of Illinois, Skoog synthesized and tested hundreds of compounds for cytokinin activity, and established the principles that govern relationships between plant structure and activity. These discoveries are generally held to be one of the major advances made in the plant sciences during the last 50 years. Skoog was also a pioneer in investigating how to control the formation of roots, stems and leaves from undifferentiated cells in plant tissue cultures. Tissues could be made to develop as undifferentiated masses of cells, or to become roots, stems or a combination of roots and stems, resulting in complete plants. Skoog and Miller (1957) demonstrated for the first time that a ratio of auxin and cytokinin can control the root, shoot and callus formation. Skoog's theory that plant development is controlled by hormone levels and other factors led to the realization that whole plants can be generated from cultured cells. This laid groundwork for the production of transgenic plants and other advances in biotechnology.

In 1962 he and graduate student Toshio Murashige published a culture medium for optimal plant tissue growth that remains in use today. This is a high salt, medium, which was modified by Linsmaier and Skoog (1965) and relatively low salt formulation was evolved. Skoog was born in Holland, Sweden on July 15, 1908. He decided to take up residence in the United States during a visit to California in 1925, and he became a U.S. citizen 10 years later. He received Bachelor of Science degree in chemistry from the California Institute of Technology in 1932. He earned his Ph.D. degree from that institution in 1936. While at the institute, he worked closely with prominent Plant Physiologists Kenneth Thimann and Fritz Went (known for their work and discovery of auxin).

Between 1937–1941, Skoog was a staff at Harward University, and from 1941–1944, at Johns Hopkins University. He served as a chemist and technical representative of the U.S. Army in Europe between 1944–1946.

Skoog played an active role in a number of biological societies. He served as president of the American Society of Plant Physiologists, the Society of Developmental Biology, the American Society of General Physiologists and the International Plant Growth Substances Association. He was a member of the American Academy of Arts and Sciences, and the National Academy of Sciences. He was awarded the National Medal of Science in 1991 during a ceremony at the White House.

Renowned Plant Physiologist and National Medal of Science recipient Folke Kari Skoog, professor of botany at the University of Wisconsin, Madison, U.S.A. for 32 years, died on Feb. 15, 2001 after a long illness He was 92.

E.C. COCKING

Professor Edward Cocking, a Fellow of Royal Society of London, works at the Centre for crop nitrogen fixation in the words major non–legume crops (especially rice, wheat, maize sorghum, and oilseed rape). This involves basic research on the interaction of crops with azorhizobia for the establishment of endophytic nitrogen fixation.

Professor Cocking and his associates have been instrumental in developing many techniques of plant cell and protoplasts culture. In 1960s, for the first time a method of isolation of protoplasts in large quantities was developed using enzymes obtained from *Myrothecium* fungus. They obtained protoplasts from root tips for *Lycopersicon esculentum* using cellulose produced by the fungus. Protoplasts isolation and their use in somatic hybridization opened new avenues of improving plant species. Subsequently, the technique found with applications because of availability of cellulose at commercial level. At that time they started attempting fusion of legume cells containing bacteria with non–legume cells towards developing non–legume nitrogen fixing plants. They continue these efforts with the establishment of research centre for nitrogen fixation studies.

They (Bhojwani and Cocking, 1972) were also successful in isolating protoplasts from pollen grains and pollen mother cells. They used protoplasts for understanding tobacco mosaic virus infection and multiplication in plant cells. Development of technique of protoplasts isolation has many applications in cell biology, plant physiology and genetic engineering as mentioned in the chapter on protoplast culture. Besides many articles on protoplasts culture, his some of the important publications are as follows:

1. Blackhall, N,W., Jotham, J.P., Azhakanandam, K., Power, J.B., Lowe, K.C., Cocking, E.C. and Davery, M.R. (1999). Callus initiation, maintenance and induction in rice *In*: Methods in Molecular Biology. Vol. III (Ed: Hall, RD) Humana Press Inc, Totowa NJ, pp. 19–29.
2. Azhakanandam, K., Mc Cabe, M.S., Lowe, K.C., Cocking, E.C., Power, J.B., and Davery, M.R. (1999). Gene transfer into and molecular characterisation of TO, T1 and T2 transgenic rice plants mediated by *Agrobacterium tumefaciens. Journal of Experimental Botany*, 50(P2.41): 20.
3. Kennedy, I.R. and Cocking, E.C. (1997). Biological nitrogen fixation; The global challenge and future needs. Position Paper of the meeting at the Rockefeller Foundation Bellagio Conference Centre, 8–12 April, 1997.

TOSHIO MURASHIGE

Professor Toshio Murashige is working as a Professor of Plant Sciences at the University of California, Riverside, U.S.A. As a graduate student he worked with Professor skoog on nutrition of plant cells using tobacco pith cells as an experimental material. During that period plant tissue culture was in its development stage Only White's medium was known. They cultured tobacco pith cells on White's medium and evaluated, one by one, inorganic and organic nutrients and formulated a medium, known as Murashige and Skoog's medium (Murashige T. and Skoog, F. 1962. A revised medium for rapid growth and bioassay with tobacco tissue culture, Physiologia Plantarum 15:473–497). In this study, the optimal concentration of each individual element was determined empirically. This is a synthetic medium which is designed to assay the effect of organic supplements. This is the most widely used medium for plant tissue culture of higher plants. Later on, several other media were developed based on this formulation.

After joining the University of California, he worked on developing micro propagation technique and established three stage micropropagation system. These are: 1. Establishment of aseptic culture, 2. Multiplication of propagule, and 3. preparation for re–establishment of plants in soil. According to him each stage has a different nutrient and light requirement. He worked on multiplication of several plant species belonging to diverse taxa *e.g.*, ornamentals like *Dracaena, Scindapcus, Syngonium.* He also worked on process of somatic embryo formation using carrot and citrus plants and described changes at molecular levels during the process.

The following two contributions are well appreciated among several hundred articles he has published.

1. Murashige, T. (1974). Plant propagation through tissue culture. *Annual Review of Plant Physiology*, 25: 134–66.
2. Murashige, T. (1978). The impact of plant tissue culture on agriculture. TA Thorpe (*Ed*) IPTCA, Calgary, Canada.

Presently, he is working as emeritus scientist on problem of rejuvenation in adult tree explants. Rejuvenation of trees is evident by restored organogenesis, leaf morphology and shoot vigor, and diminishing leaf abscission, chlorosis and tissue, and culture medium discoloration *in vitro*. Investigation of the underlying mechanism, using *in vitro* growing shoots, disclosed identical agarose gel electrophoretic patterns of mitochondrial DNA extracted from juvenile and rejuvenated shoots and distinct from the mitochondrial DNA of adult shoots.

Huang, L.C., Hsaio, C.K., Lius, S., Liu, S.F., Huang, B.L. and Murashige, T. (1992). Restoration of vigor and rooting competence in stem tissues of mature citrus by repeated grafting of shoot apices onto freshly germinated seedings *in vitro. In vitro Cell. Dev. Biology–Plant,* 28: 30–32.

P. MAHESHWARI

The most outstanding Indian embryologist, P. Maheshwari was trained at Allahabad by Dudgeon. Maheshwari remained as expert of Botany for almost three decades. After completing his D.Sc. Thesis (1929) at Allahabad, he moved to the Agra College (1931–1937), and after serving brief tenures at Allahabad (1937–1939), Lucknow (1939) and Dacca (1939–1949), at the invitation of the Vice–Chancellor (Sir Maurice Gwyier), he joined the University of Delhi (1949–1966) and established a very productive school of embryology. The earlier work of Prof. Maheshwari was concerned with embryology, which he started in 1930 at Agra. By 1933, a flourishing school of Angiosperm Embryology was established and in coming years, this centre became internationally known.

B.M. Johri, V. Puri, B. Singh and H.R. Bhargava were the prominent students, trained at the above centre. Johri's work on the Alismaceae and Butomaceae done between 1933 and 1938 is among the first important piece of research in embryology in the country. Occurrence of pollen grains in the stylar canal of a typical *angiosperm–Butompopsis lanceolata* by Johri (1936) was an appreciable discovery. P. Maheshwari worked at Haward University in 1945 and devoted most of his time to complete the book entitled, "An introduction to the embryology of angiosperms" (1950). He had also edited a few symposium volumes in 1962–63. He also established the International Society of Plant Morphologists and started an international journal "Phytomorphology." It was due to continuous efforts of P. Maheshwari that India acquired an international status in embryology. In 1965, he was elected fellow of the Royal Society of London.

In India, studies on tissue culture started in mid 1950 at the Department of Botany, University of Delhi under the directions of Professor P. Maheshwari. This was beginning of experimental embryology. Some faculty members who were trained abroad to acquire knowledge were–S. Narayanswami who worked with C.D. La. Rue at Michigan, and B.M. Johri who worked with H.E. Street at Swansea.

Over the past four decades no other institution in India has made more significant contributions than the Department of Botany, Univers.ity of Delhi, in the use of tissue culture methodologies for morphogenetic studies involving ovary, ovule, embryo, endosperm, cotyledons, and other reproductive organs. These works deal with isolation, culture, *in vitro* growth of ovary, ovule, and embryo parts in fruit set and seed development in response to exogenous growth regulators has been achieved. Emphasis was also laid on the *nucellar* culture for obtaining true clones of *Citrus* and the endosperm culture for the production of triploid plant ih case of *Citrus, Oryza* and *Putranjiva roxburghii.*

S. GUHA MUKHERJEE AND S.C. MAHESHWARI

Plant tissue culture studies lead to the demonstration for the first time of the development of haploids through anther and pollen culture by Sipra Guha Mukherjee and S.C. Maheshwari, (1964–67). This research finds many applications in crop and tree breeding programmes, being, a quicker and simpler method of producing homozygous lines than conventional breeding programme. Professor S.C. Maheshwari was working in the Botany Department of Delhi University. Later on he joined the newly formed Department of

Molecular Biology of Delhi University at South Campus. He contributed significantly in the development of these departments in the University. The university Grants Commission and Department of Science and Technology have recognized the research credential of Prof. Maheshwari and granted him a centre for research on molecular biology. During his long research career, he worked on various aspects of plant growth and developments like protoplasts culture, characterization of chloroplast DNA, regeneration in cereals, gene isolation and sequencing, northern analysis, transcript mapping, genomic mapping and gene delivery using electroporation in plants like spinach, mung bean, and rice towards better understanding the mechanism of gene expression during development. Presently he is working in International Centre for Genetic Engineering and Biotechnology at New Delhi.

Prof. Sipra Guha Mukherjee is presently working in the School of Life Sciences, J.N. University, at New Delhi. She is working on regeneration of plants and mechanism of regeneration involving various enzymes, membrane phospholipids and second messengers.

PROFESSOR SANT SARAN BHOJWANI

Sant Saran Bhojwani was born to Mrs. Nam Adhari and Mr. Parmanand on 20th November, 1940 in the serene and tranquil environment of Dayalbagh, about 3 km from the hustle–bustle of the Agra city. He had his early education in Dayalbagh and graduated and postgraduated from Agra University. Soon after finishing M.Sc. (Botany), Dr Bhojwani served his alma–mater (R.E.I. Dayalbagh, Agra) as lecturer for one year before joining the University of Delhi as a doctoral student. His supervisor, late Professor B.M. Johri assigned him a challenging research problem, with the warning that his Ph.D. degree would depend on his demonstrating the cellular totipotency of endosperm, a completely unorganized, short–lived, triploid tissue. Earlier, many students of Professor Johri and scientists elsewhere in the world could establish tissue cultures of endosperm but failed to induce the organogenic differentiation. It was remarkable that within six months of his joining Delhi University, Dr. Bhojwani achieved differentiation of normal shoot buds from the endosperm of *Exocarpus cupressiformis*; a parasitic flowering plant (*Nature*, 1965). At this stage a very renowned American Plant Physiologist, Prof. F.C. Steward, visited the University of Delhi who even after observing the cultures could not believe that the endosperm tissue could form shoots and remarked, "Young man, take a bet with me. All the shoots in the cultures are diploid. If that is the case remember me or else forget me". However, when the shoots of endosperm origin were cytologically analysed, all of them were found to be triploid, which is of considerable practical importance in plant genetics and improvement. Subsequently, Dr Bhojwani established the cellular totipotency of endosperm cells by reporting regeneration of triploid shoots and/or plants in *Scurrula pulverulenta, Acacia nilotica* (Garg *et al*., 1996), *Morus alba* (Thoma *et al*., 2000) and *Azadirachta indica* (Chaturvedi *et al*., 2003). In the meantime, many other scientists confirmed the observations of Bhojwani.

Dr. Bhojwani and his students worked on a range of basic and applied aspects of *in vitro* plant morphogenesis. During 1971–1972 he worked with Dr Norman Sunderland at the John Innes Institute, Norwich, U.K. under the British Council Fellowship Programme and reported quantitative changes in nucleic acid and protein contents of microspores during the induction of androgenesis in tobacco using histochemistry and cytophotometry (*J. Exp. Bot.* 1973). In 1972, he spent three months in the laboratory of Professor Edward C. Cocking, FRS, at the University of Nottingham, U.K. and reported for the first time isolation of microspore protoplasts using helicase enzyme. The report appeared in *Nature, New Biology* (1972). Dr Bhojwani had another opportunity to work in the U.K. for a year during 1975–1976 under the Royal Society Commonwealth Bursary. Fortunately, a major research project on "Micropropagation of Important Horticultural and Silvicultural species of India" was sanctioned to him by the UGC, under which he and his students developed an efficient protocol for clonal propagation of the leguminous tree species, *Leucaena leucocephala,* and *in vitro* nodulation of micropropagated plants by *Rhizobium* to enhance their field survival. He also demonstrated that sugar cubes, produced by Daurala Sugar Mills, was a fair substitute of 'Anal R' Grade Sugar used in Plant tissue culture media. The sugar cubes were more than 10 times cheaper than the 'Anal R' Grade sucrose. In 1985 the Department of Environment and Forests, Government of India, awarded another major research project to Dr Bhojwani to work on "*In vitro* Conservation of Endangered Plants". It led to the development of protocols for micropropagation and cold storage of Himalayan Species of three medicinally important plants, *viz*., *Picrorhiza kurroa*, *Podophyllum hexandrum* and *Saussurea lappa*. In collaboration with the scientists at the Biochemical Engineering and Biotechnology Department of IIT Delhi, Dr Bhojwani studied the kinetics of cell growth in suspension cultures of *Podophyllum hexandrum* and *in vitro* production of Podophyllotoxin, an anticancerous drug (*Biotechnol. Lett.* 2001, *J. Biosci. Bioengg.*, 2002). Dr Bhojwani guided six Ph.D.'s on plant regeneration alone from somatic and gametic cells of *Brassica* spp and published several papers (*Plant Cell Tissue Organ Cult.* 1985, 1991; *Biol. Plant.*, 1989; *Plant Sci.* 1990a, b; *Euphytica* 1993). His group also achieved high frequency androgenesis and selection of agronomically useful androclones in *B. juncea*. This work was supported by funds from MOMBUSHO, Japan and European Commission, Brussels. Dr Bhojwani undertook two major projects on mulberry biotechnology and investigated micropropagation of some elite clones and production of gynogenic haploids (*Euphytica*, 1999) and endosperm derived triploids (*Pant Cell Rep.* 2000) of this invaluable tree for silk industry, the sole source of feed for silkworms. Recently, he has reported the production of gynogenic haploids (*Plant Cell Rep.* 2003) and triploids (*J. Plant Physiol.*, 2003) of Neem. Dr Bhojwani has published 75 research papers in journals of international repute, 10 critical reviews and 19 invited chapters in books published from India and abroad. In addition, he has authored and edited several books. His first book "The Embryology of Angiosperms" (Vikas Publishers, New Delhi) has been a popular text book for graduate and post–graduate students in India and many other countries. Running into its 5th edition, the book has been translated into Japanese (1995) and Korean (2001). In 1983, Dr Bhojwani brought out another book titled "Plant Tissue Culture: Theory and Practice", published by Elsevier, The Netherlands. This has been regarded as the first standard text book on the subject and became so popular worldwide that the publishers brought out its paperback edition in 1986. It was translated into Korean in 1986. Under a project funded by the Department of Biotechnology, Dr Bhojwani completed a mammoth task of compiling 'A Classified Bibliography of Plant Tissue Culture', covering the entire literature on the subject up to 1984. He spent two weeks in the U.K. under the INSA–Royal Society Exchange Programme to complete the volume. It soon became a popular reference book. Dr. Bhojwani has edited four volumes, *viz.*, "Plant Tissue Culture: Applications and Limitations" (1990; Elsevier), "Morphogenesis in Plant Tissue Cultures" (1999; Kluwer Academic Publishers, The Netherlands), PROFESSOR SANT SARAN BHOJWANI vii "Current Trends in the Embryology of Angiosperms" (2002; Kluwer Academic Publishers) and "Agrobiotechnology and Plant Tissue Culture" (2003; Science Publishers, U.S.A.).

Dr Bhojwani has been in great demand by the organisers of conferences, seminars, workshops, training courses and refresher courses because of his contributions and in–depth knowledge in the subject of Plant Tissue Culture. Dr. Bhojwani has participated in several National and International Conferences in India and overseas. He was invited to deliver a lecture at the conference on "Problems Related to Mass Propagation of Horticultural Species", Belgium (1985). The organisers of the Conference on Tissue Culture of Tropical Plants in Bagota, Colombia invited him to deliver a Plenary Lecture and Chair a session. Dr Bhojwani was a member of the International Advisory Committee of the VIII Conference of the International Association of Plant Tissue Culture held in Florence, Italy, where he organized a workshop. He was also a member of the International Advisory Committees of the 1st, 2nd and 3rd Asia–Pacific Conferences in Taejon, South Korea (1993), Shanghai, China (1997) and Singapore (2000). He delivered plenary lectures in the Conferences held in South Korea and Singapore. In 1987 an International Symposium on Gene Manipulation for Plant Improvement was organized in Kuala Lumpur, Malaysia and Dr Bhojwani was invited to deliver a plenary lecture. Dr. Bhojwani also attended the VI Conference of the International Association of Plant Tissue Culture held in Minnesota, U.S.A. (1986) and the International Botanical Congress in Yokohama, Japan (1993). He also delivered an invited talk in the latter. He was the only Indian invited as a Resource Person to a workshop on "Production and Utilization of Double–Haploid Lines in Rice Breeding" organized by the International Agency for Atomic Energy in Suwon, South Korea, in 1999. Dr. Bhojwani delivered plenary lectures and chaired sessions in International Conferences in Dhaka, Bangladesh. Recently, he was invited to participate in the 15th Biennial Conference of the New Zealand Chapter of International Association of Plant Tissue Culture and Biotechnology" at Leigh, New Zealand and presented a paper on "Pollen Embryogenesis in *Brassica* ssp. Dr. Bhojwani has been a recipient of many honours. He was elected Full Member of the New Zealand Institute of Agricultural Sciences (1981). In 1990 he became Invited Member of the Technology Transfer Association of Japan. He was awarded the Nawashina Memorial Medal. Dr Bhojwani was elected as a Fellow of the National Academy, Allahabad in 1994 and has been awarded many National and International Fellowships to visit laboratories in other countries. Besides the British Council Fellowship and Royal Society Bursary to work in U.K., Dr. Bhojwani was awarded the Senior Fellowship of the NRAC, New Zealand; Fellowship of the Japanese Society for the Promotion of Science; Biotechnology Overseas Associateship, Government of India; CIDA/ NSERC Research Associateship, Canada; INSA–KOSEF Fellowship of South Korea, and Fellowship of the Kernfor-schungsanlage, Germany. Dr. Bhojwani has been on the Editorial Board of many journals. To mention a few, *Scientia Horticulture,* Holland; *Journal of Biochemistry and Biotechnology*, New Delhi; *Phytomorphology*, Delhi, *Plant Tissue Culture*, Dhaka and *Chromosome*, Calcutta. Dr Bhojwani has been a member of the Academic Council's of the TERI School of Advanced Study and C.C. Singh University, Meerut. He is the Chairman of the Research Advisory Committee viii PROFESSOR SANT SARAN BHOJWANI of the Central Tassar Research and Training Institute, Ranchi. He was a Visiting Senior Fellow of the Tata Energy Research Institute, New Delhi and made a major contribution to the designing and production and planning of the DBT–Sponsored Plant Tissue Culture Pilot Plant. After serving the University of Delhi for 35 years, Professor Bhojwani took voluntary retirement to serve the Dayalbagh Educational Institute (Deemed University), Agra as its Honorary Director. Married to Shaku, Dr. Bhojwani discharged his family obligations well and timely with both the children married and settled happily with their families.

DR. M.K. RAJDAN

Dr. M.K. Razdan is the Principal of Shyam Lal College, University of Delhi. Prior to this he was Vice Principal at Ramjas College of the same University, were he taught for more than three decades. Besides teaching students at undergraduate and postgraduate level Dr. Razdan was Head of the Botany Department at Ramjas College

during the period 1976-1998 and subsequently Coordinator of add-on courses on 'Biotechnology' as well as 'Environment Management and Law' in the same Institute. Dr. M.K. Rajdan has been teaching at a constituent college of the university of Delhi for more than two decades now. He has also been a visiting scientist at USDA, Vegetable Labora-tory, Belsville, Maryland and Department of natural Resource and science and Landscape Architecture, University of Maryland, USA during the period 1999-2001. Dr. Raazdan, was awarded a prestigious commonwealth Bursary by the Royal Society, London, which enable him to work on aspect of plant somatic cell genetics at the Department of Botany, University of Nottingham, U.K., during 1978–79. Beside publication in Reputed research journals, Dr. Razdan has authored several books, *viz.* (1) Plant Tissue Culture: Theory and Practice, Elsevier, Amsterdam, 1983, Revised Edition 1996 (S.S. Bhojwani and M. K. Razdan), and (3) Conservation of Plant Genetic Resources, Vols. 1 and 2 (M. K. Razdan and E.C. Cocking), Science Publishers, USA (1997, 2000). He has participated in various symposia and congresses on plant cell/tissue culture. During the period of 1982–90 he was joint Editor and Member of the Editorial Board of the Journal of Cytology and Genetics. He was elected Fellow of the Society of Cytologist and Geneticists (FSCG) in December 1988 and visited Moscow, in 1993, to deliver lecturers as well as to exchange scientific experiences at various institute of Russian Academy of Sciences under cultural exchange programme sponsored by the University Grants Commission, India. He participated at International Conference/Workshop/Seminars as delegate, organizer and resource person. He is currently Member of Delhi University Academic Council and Member of Delhi University Court. He has also been examiner for evaluation of Ph.D Thesis from other universities and referee of Research Papers sent for publication in various journals.

Glossary

Absorption : In general: the process of absorbing; taking up of water and nutrients by assimilation or imbibition. The taking up by capillary, osmotic, chemical or solvent action, such as the taking up of a gas by a solid or liquid, or taking up of a liquid by a solid. *cf* adsorption.

Abiotic Stress : The effect of non–living factors which can harm living organisms. These non–living factors include drought, extreme temperatures, pollutants, etc.

Abscisic Acid : A plant growth regulator involved in abscission, dormancy, stomatal opening/closure, and inhibition of seed germination. It also affects the regulation of somatic cell embryogenesis in some plant species.

Accessory Bud : Lateral bud occurring at the base of a terminal bud or at the side of an axillary bud.

Acclimatization : The adaptation of a living organism (plant, animal or micro–organism) to a changed environment that subjects it to physiological stress.

Acellular : Describing tissues or organisms that are not made up of separate cells but often have more than one nucleus. *cf* syncytium.

Adventitious : Developing from unusual points of origin, such as shoots or root tissues from callus or embryos from sources other than zygotes. This term can also be used to describe agents which contaminate cell cultures.

Aflatoxin : A potent toxin produced by certain fungi that commonly grow on damp or poorly dried seeds, grains and nuts.

Agar : A polysaccharide powder derived from algae used to gel a medium. Agar is generally used at a concentration of 6–12 g/liter.

Anchorage : Dependent Cells or Cultures– Cells, or cultures derived from them, which will grow, survive, or maintain function only when attached to a surface such as glass or plastic. The use of this term does not imply that the cells are normal or that they are or are not neoplastically transformed.

Androgenesis : Development of plants from male gametophytes.

Aneuploid : The situation which exists when the nucleus of a cell does not contain an exact multiple of the haploid number of chromosomes; one or more chromosomes being present in greater or lesser number than the rest. The chromosomes may or may not show rearrangements.

Anther : The terminal portion of a stamen which contains pollen in pollen sacs.

Anther Culture : Culture of excised anthers, to obtain haploids.

Anthesis : The flowering period or efflorescence. Anthesis is the time of full bloom, which lasts till fruit set.

Anthocyanin : Water–soluble blue, purple and red flavonoid pigments found in vacuoles of cells.

Antiauxin : A chemical that interferes with the auxin response. Antiauxin may or may not involve prevention of auxin transport or movement in plants. Some antiauxins are said to promote morphogenesis *in vitro,* such as 2,3,5–tri–iodobenzoate (TIBA; f.w. 499.81), or 2,4,5–trichlorophenoxyacetate (2,4,5–T; m.w. 255.49), which stimulate the growth of some cultures.

Antioxidant : Compound that slows the rate of oxidation reactions.

Apical : Pertaining to tip or apex.

Apical Dominance : The phenomenon of suppression of growth of an axillary bud in the presence of the terminal bud on the branch.

Apical Meristem : A region of the tip of each shoot and root of a plant in which cell division is continually occurring to produce new stem and root tissue, respectively. Two regions are visible in the apical meristem: (i) An outer 1–4–cell layered region (called the tunica), where cell divisions are anticlinal, *i.e.*, perpendicular to the surface; and below the tunica, (ii) the corpus, where the cells divide in all directions, giving them an increase in volume.

Apoenzyme : Inactive enzyme that has to be associated with a specific organic molecule called a co–enzyme in order

	to function. The apoenzyme/co–enzyme complex is called a holoenzyme.
Apomixis (Gr. *apo*, away from + *mixis*, a mingling; adj: apomictic) :	The asexual production of diploid offspring without the fusion of gametes. The embryo develops by mitotic division of the maternal or paternal gamete, or, in the case of plants, by mitotic division of a diploid cell of the ovule. *cf* androgenesis; gynogenesis; parthenogenesis.
Asepsis :	Without infection or contaminating microorganisms.
Aseptic Culture :	A culture without undesired of foreign life form. An aseptic culture means the purposeful culture of different types of cells, tissues of higher plant without any contamination of bacteria, fungi etc.
Aseptic Technique :	Procedures used to prevent the introduction of fungi, bacteria, viruses, mycoplasma or other microorganisms into cell, tissue and organ culture. Although these procedures are used to prevent microbial contamination of cultures, they also prevent cross contamination of cell cultures as well. These procedures may or may not exclude the introduction of infectious molecules.
Asexual :	Not involving sex.
Attachment Efficiency :	The percentage of cells plated (seeded, inoculated) which attach to the surface of the culture vessel within a specified period of time. The conditions under which such a determination is made should always be stated.
Auxin :	A group of growth substances produced in the zone of elongation that cause cell elongation, apical dominance, root initiation etc. 2,4–D (2,4– dichlorophenoxyacetic acid), IAA (Indole–3–acetic acid), NAA (á–naphaleneacetic acid), IBA (Indole butyric acid) are some of the synthetic auxines commonly used in plant tissue culture.
Autoclave :	A machine capable of sterilizing wet or dry items with steam under pressure.
Axenic Culture :	A culture without foreign or undesired life forms. An axenic culture may include the purposeful cocultivation of different types of cells, tissues or organisms.
Axillary Bud :	A bud found at the axil of a leaf (synonymous with lateral bud).
Axillary Bud Proliferation :	Propagation in culture using protocols and media which promote axillary (lateral shoot) growth. This is

a technique for mass production (micropropagation) of plantlets in culture, achieved primarily through hormonal inhibition of apical dominance and stimulation of lateral branching.

Axillary Shoot : Shoot that develops from the bud located in the axil of a leaf.

Backbulb : An old pseudobulb, often without leaves, but still alive and bearing one or more buds behind the actively growing portion of a sympodial orchid.

Batch Culture : A type of cell suspension culture in which cell or small cell aggregates grow in a definite volume of nutrient medium.

Bio–Conversion : The transformation and breakdown of substances from one form to another or to its constituents through the activities of micro organisms or enzyme.

Bio–Enrichment : Adding nutrients or oxygen to increase microbial breakdown of pollutants.

Bioreactor : A vessel used for large scale aseptic culture of cells, embryos or shoots under controlled growth condition.

Biosynthesis : Formation of chemicals by living cells.

Biotechnology : Application of technique to modify the fundamental process of growth and reproduction in plants, animals or microorganism to enhance productivity or to produce new product.

Browing : Used here to describe the occasional purplish/ darkening of media by oxidized product of phenol leached out of explant which may be phytotoxic.

Callus Tissue : An unorganized mass of proliferative cells produced either in culture or in nature.

Carbohydrates : Organic compounds, composed of carbon, hydrogen and oxygen such as sugars, starch and cellu lose.

Catalyst : A substance that affect the rate of a reaction with out otherwise being involved in that reaction.

Cell Culture : Term used to define the maintenance or culture of cells *in vitro* including the culture of single cell. In culture the cells are no longer organized in to tissues.

Cell Generation Time : The interval between consecutive divisions of a cell.

Cell Hybridization : The fusion of two or more dissimilar cells leading to the formation of a synkaryon.

Cell Line : A cell line arises from a primary culture at the time of the first successful subculture. The term "cell line" implies that cultures from it consist of lineages of cells originally present in the primary culture. The terms finite or continuous are used as prefixes if the status of the culture is known. If not, the term line will suffice. The term "continuous line" replaces the term "established line". In any published description of a culture, one must make every attempt to publish the characterization or history of the culture. If such has already been published, a reference to the original publication must be made. In obtaining a culture, as originally named and described, must be maintained and any deviations in cultivation from the original must be reported in any publication.

Cell Strain : A cell strain is derived either from a primary culture or a cell line by the selection or cloning of cells having specific properties or markers. In describing a cell strain, its specific features must be defined. The terms finite or continuous are to be used as prefixes if the status of the culture is known. If not, the term strain will suffice. In any published description of a cell strain, one must make every attempt to publish the characterization or history of the strain. If such has already been published, a reference to the original publication must be made. In obtaining a culture from another laboratory, the proper designation of the culture, as originally named and described, must be maintained and any deviations in cultivation from the original must be reported in any publication.

Chemostat : A type of cell suspension culture in which growth rate and cell density remain constant by fixed rate of input of growth limiting nutrient.

Chimera : Organism and tissue that is not genetically homogeneous.

Clone : A group of genetically identical individuals produced from a single explant.

Chemically Defined Medium : A nutritive solution for culturing cells in which each component is specifiable and ideally, is of known chemical structure.

Clonal Propagation : Asexual reproduction of plants that are considered to be genetically uniform and originated from a single individual or explant.

Cloning Efficiency : The percentage of cells plated (seeded, inoculated) that form a clone. One must be certain that the colonies formed arose from single cells in order to properly use this term.

Closed Continuous Culture : A continuous culture in which in flow of fresh medium is balanced by outflow of corresponding volumes of spent medium. Cells are separated mechanically from outflowing medium and added back to the culture.

Coconut Milk/ Water : The liquid endosperm from immature coconuts, used as a supplement in the culture medium.

Colony Forming Efficiency : The percentage of cells plated (seeded, inoculated) that form a colony.

Complementation : The ability of two different genetic defects to compensate for one another.

Contact Inhibition of Locomotion : A phenomenon characterizing certain cells in which two cells meet, locomotory activity diminishes, and the forward motion of one cell over the surface of the other is stopped.

Contamination : Being infested with unwanted microorganisms such as bacteria or fungi.

Continuous Cell Culture : A culture which is apparently capable of an unlimited number of population doublings; often referred to an as immortal cell culture. Such cells may or may not express the characteristics of *in vitro* neoplastic or malignant transformation.

Crisis : A stage of the *in vitro* transformation of cells. It is characterized by reduced proliferation of the culture, abnormal mitotic figures, detachment of cells from the culture substrate, and the formation of multinucleated or giant cells. During this massive cultural degeneration, a small number of colonies usually, but not always, survive and give rise to a culture with an apparent unlimited *in vitro* lifespan. This process was first described in human cells following infection with an oncogenic virus (SV40).

Cryopreservation : Ultra–low temperature storage of cells, tissues, embryos or seeds. This storage is usually carried out using temperatures below –100°C.

Culture : A plant growing *in vitro*.

Cybrid : The viable cell resulting from the fusion of a cytoplast with a whole cell, thus creating a cytoplasmic hybrid.

Cytokinin : A group of plant growth regulators that regulate growth and morphogenesis and stimulate cell division. Endogenous cytokinins, cytokinins that occur naturally, include zeatin and 6–ã, ã–dimethylally-laminopurine (2iP). Exogenous cytokinins, cytokinins that are man–made or synthetic, include 6–furfuryl-aminopurine (kinetin) and 6–benzylaminopurine (BA or BAP).

Cytoplast : The intact cytoplasm remaining following the enucleation of a cell.

Cytoplasmic hybrid : Synonymous with "cybrid"

Cytoplasmic inheritance : Inheritance attributable to extranuclear genes; for example genes in cytoplasmic organelles such as mitochondria or chloroplasts, or in plasmids, etc.

De differentiation : Reversal of organized structure into an undifferentiated state.

Deionize : Reversal of organized structure into an undifferentiated state.

Density : Dependent Inhibition of Growth–Mitotic inhibition correlated with increased cell density.

Differentiated : Cells that maintain, in culture, all or much of the specialized structure and function typical of the cell type in vivo.

Diploid : The state of the cell in which all chromosomes, except sex chromosomes, are two in number and are structurally identical with those of the species from which the culture was derived. Where there is a Commission Report available, the experimenter should adhere to the convention for reporting the karyotype of the donor. Commission Reports have been published for mouse, human, and rat. In defining a diploid culture, one should present a graph depicting the chromosome number distribution leading to the modal number determination along with representative karyotypes.

Disease Free : A plant certified through specific tests as being free of specified pathogens.

Dominant Gene/ Character : One that is expressed in a diploid cell or organism irrespective of its allel present on the homologous.

Downstream Processing : Bioprocessing steps following fermentation and/or cell culture, a sequence of separation and purification activities needed to obtain the required drug product at the necessary level of purity.

Electroporation : Creation, by means of an electrical current, of transient pores in the plasmalemma usually for the purpose of

		introducing exogenous material, especially DNA, from the medium.
ELISA Serological Method	:	A method used to test for presence of viruses in plant tissues. The acronym ELISA stands for Enzyme Linked Immunosorbent Assay.
Embryo	:	An organized structure formed following a pre–determined mode of development inside the female gametophyte with or without fertilization.
Embryo Culture	:	*In vitro* development or maintenance of isolated mature or immature embryos.
Embryogenesis	:	The process of embryo initiation and development.
Endosperm	:	The nutritive tissue or liquid surrounding the developing embryo in seeds.
Epigenetic Event	:	Any change in a phenotype which does not result from an alteration in DNA sequence. This change may be stable and heritable and includes alteration in DNA methylation, transcriptional activation, translational control and posttranslational modifications.
Epigenetic Variation	:	Phenotypic variability which has a nongenetic basis.
Epithelial–Like	:	Resembling or characteristic of, having the form or appearance of epithelial cells. In order to define a cell as an epithelial cell, it must possess characteristics typical of epithelial cells. Often one can be certain of the histologic origin and/or function of the cells placed into culture and, under these conditions, one can be reasonably confident in designating the cells as epithelial. It is incumbent upon the individual reporting on such cells to use as many parameters as possible in assigning this term to a culture. Until such time as a rigorous definition is possible, it would be most correct to use the term "epithelial–like".
Etiolated	:	A state of growth induced.
Euploid	:	The situation which exists when the nucleus of a cell contains exact multiples of the haploid number of chromosomes.
Explant	:	Tissue taken from its original site and transferred to an artificial medium for growth or maintenance.
Explant Culture	:	The maintenance or growth of an explant in culture.
Exudate	:	Matter discharged or diffused from cultured tissue.
Fed–Batch Fermentation	:	Culture of cells or micro–organisms where nutrients are added periodically to the bioreactor.

Feedback Inhibition : The process by which the accumulated end product of a biochemical pathway stops synthesis of that product. A late metabolite of a synthetic pathway regulates synthesis at an earlier step of the pathway. *cf* end–product inhibition.

Fermentation Substrates : Materials used as food for growing micro–organism. The fermentation substrates and the trace materials needed, together with chemicals added to make the fermentation easier, form the culture medium.

Feeder Layer : A layer of cells (usually lethally irradiated for animal cell culture) upon which are cultured a fastidious cell type.

Fibroblast–Like : Resembling or characteristic of, having the form or appearance of fibroblast cells. In order to define a cell as a fibroblast cell, it must possess characteristics typical of fibroblast cells. Often one can be certain of the histologic origin and/or function of the cells placed into culture and, under these conditions, one can be reasonably confident in designating the cells as fibroblast. It is incumbent upon the individual reporting on such cells to use as many parameters as possible in assigning this term to a culture. Until such time as a rigorous definition is possible, it would be most correct to use the term "fibroblast–like."

Finite Cell Culture : A culture which is capable of only a limited number of population doubling after which the culture ceases proliferation.

Friability : A term indicating the tendency for plant cells to separate from one another.

Gamete : A sex cells derived from a male or female gametophyte, with one half number of chromosome present in the somatic cells of the plant.

Gametoclonal Variation : Variation in phenotype, either genetic or epigenetic in origin, expressed by gametoclones.

Gametoclone : Plants regenerated from cell cultures derived from meiospores, gametes or gametophytes.

Gene : One of a set of units of heredity, having a specific effect on the characteristics of an organism. Genes are composed of DNA and are arranged linearly on the chromosomes.

Genetic engineering : A laboratory technique for genetic modification of plants by introducing one or a few selected genes into cells to change their inherited characteristics.

Growth Regulator : Organic compounds other than nutrients that, in small amount, influence growth, differentiation and multiplication, such as auxins, cytokinin, gibberellin and ethylene.

Gibberellins : A plant growth regulator that influences cell enlargement. Endogenous growth forms of gibberellin include Gibberellic Acid (GA_3).

Habituation : The acquired ability of a population of cells to grow and divide independently of exogenously supplied growth regulators.

Haploid : Having single copy per cell of each chromosome characteristic of the species (=n).

HEPA Filter : High efficiency particulate air filters, an essential component of laminar air flow hood used for aseptic manipulation of tissue culture.

Heterokaryon : A cell possessing two or more genetically different nuclei in a common cytoplasm, usually derived as a result of cell–to–cell fusion.

Heteroploid : The term given to a cell culture when the cells comprising the culture possess nuclei containing chromosome numbers other than the diploid number. This is a term used only to describe a culture and is not used to describe individual cells. Thus, a heteroploid culture would be one which contains aneuploid cells.

Histiotypic : The *in vitro* resemblance of cells in culture to a tissue in form or function or both. For example, a suspension of fibroblast–like cells may secrete a glycosamino-glycan–collagen matrix and the result is a structure resembling fibrous connective tissue, which is, therefore, histiotypic. This term is not meant to be used along with the word "culture". Thus, a tissue culture system demonstrating form and function typical of cells in vivo would be said to be histiotypic.

Homokaryon : A cell possessing two or more genetically identical nuclei in a common cytoplasm, derived as a result of cell–to–cell fusion.

Hardening off : The process of gradual acclimatization of tissue culture raised plant to green house or field condition such as low humidity, high light intensity and autotrophic mode of nutrition.

Homozygous : Diploid or polyploid individuals having identical alleles on the homologues chromosomes. Self-fertilization of homozygous individuals would give a homogeneous population.

Hormones : Natural or synthetic chemicals that strongly affect growth, development or metabolism at very law concentration, *i.e.*, cytokinins, auxins, and gibberellins.

Horizontal Laminar Flow Unit : An enclosed work area that has sterile air moving across it. The air moves with uniform velocity along parallel flow lines. Room air is pulled into the unit and forced through a HEPA (High Energy Particulate Air) filter, which removes particles 0.3 µm and larger.

Hybrid cell : The term used to describe the mononucleate cell which results from the fusion of two different cells, leading to a formation of a synkaryon.

Hybridoma : The cell which results form the fusion of an antibody producing tumor cell (myeloma) and an antigenically–stimulated normal plasma cell. Such cells are constructed because they produce a single antibody directed against the antigen epitope which stimulated the plasma cell. This antibody is referred to as a monoclonal antibody.

Hyperhydration : An abnormality shown by some shoot developed in culture, in which the shoot appear as overly succulent, crisp, watered logged, or glassy tissues. Earlier term "vitrification".

Hypocotyl : The portion of a seedling stem below the cotyledons and above the roots.

Immortalization : The attainment by a finite cell culture, whether by perturbation or intrinsically, of the attributes of a continuous cell line. An immortalized cell is not necessarily one which is neoplastically or malignantly transformed.

Induction : Initiation of a structure, organ or process *in vitro*.

***In vitro* Neoplastic Transformation** : The acquisition, by cultured cells, of the property to form neoplasms, benign or malignant, when inoculated into animals. Many transformed cell populations which arise *in vitro* intrinsically or through deliberate manipulation by the investigator, produce only benign tumors which show no local invasion or metastasis following animal inoculation. If there is supporting evidence, the term "*in vitro* malignant neoplastic transformation" or "*in vitro* malignant transformation" can be used to indicate than an injected cell line does, indeed, invade or metastasize.

In vitro Propagation : Propagation of plants in a controlled, artificial environment, using plastic or glass culture vessels, aseptic techniques and a define growing medium.

In vitro Senescence : In vertebrate cell cultures, the property attributable to finite cell cultures; namely, their inability to grow beyond a finite number of population doublings. Neither invertebrate nor plant cell cultures exhibit this property.

In vitro Transformation : A heritable change, occurring in cells in culture, either intrinsically or from treatment with chemical carcinogens, oncogenic viruses, irradiation, transfection with oncogenes, etc. and leading to the acquisition of altered morphological, antigenic, neoplastic, proliferative or other properties. This expression is distinguished form "*in vitro* neoplastic transformation" in that the alterations occurring in the cell population may not always include the ability of the cells to produce tumors in appropriate hosts. The type of transformation should always be specified in any description.

In vivo : Literally 'in life' applied to any process accuring in a whole organism

Indexing : Testing of plants for pathogens or contaminants.

Inoculum : Aliquot of a suspension culture used for its sub culture

Juvenile : A phase in the sexual cycle of a plant characterized by differences in an appearance form the adult and which lacks the ability to respond to flower–inducing stimuli.

Karyoplast : A cell nucleus, obtained from the cell by enucleation, surrounded by a narrow rim of cytoplasm and a plasma membrane.

Karyotype : The characteristic number, size and shape of the chromosomes during metaphase of somatic cells of an organism.

Leading Strand : The strand of DNA that is synthesized continuously during replication.

Leaf blade : The usually flattened portion of the leaf.

Leaf bud cutting : A cutting that includes a short section of stem with attached leaf.

Leaf Margin : The edge of a leaf.

Leaf Primordium : (L. *primordium*, a beginning) A lateral outgrowth from the apical meristem, which will become a leaf when fully developed and expanded.

Leaf Roll : Virus diseases characterized by symptomatic curling of the host's leaves.

Leaf Scar : Mark left on a stem after leaf abscission.

Leaflet : Expanded leaflike part of a compound leaf.

Liposome : A closed lipid vesicle surrounding an aqueous interior; may be used to encapsulate exogenous materials for ultimate delivery of these into cells by fusion with the cell.

Membrane Bioreactors : Bioreactors where cells grow on or behind a permeable membrane, which lets the nutrients for the cell through but retains the cells themselves. A variations on this theme is the hollow–fibre reactor.

Medium : A combination of nutrients and water, contain several salts, a carbohydrate and vitamins. Such a medium, liquid or gelled, is often referred to as a basal medium and may be supplemented with growth hormones.

Meristem Culture : *In vitro* culture of a generally shiny, dome–like structure measuring less than 0.1 mm in length when excised, most often excised from the shoot apex.

Microcell : A cell fragment, containing one to a few chromosomes, which is formed by the enucleation or disruption of a micronucleated cell.

Micronucleated Cell : A cell which has been mitotically arrested and in which small groups of chromosomes function as foci for the reassembly of the nuclear membrane thus forming miconuclei the maximum of which could be equal to the total number of chromosomes.

Micropropagation : *In vitro* clonal propagation of plants form shoot tips or nodal explants, usually with an accelerated proliferation of shoots during subcultures.

Molecular weight : The sum of atomic weight in a molecule.

Morphogenesis : The evolution of a structure from an undifferentiated to a differentiated state. In other words, the process of growth and development of differentiated structures.

Mutant : A phenotypic variant resulting from a changed or new gene.

Nurse Culture : An isolated cell which generally fails to divide when plated directly on the medium used for callus cultures is able to divide under the nursing effect of the callus.

Open continuous culture : A continuous culture in which inflow of fresh medium is balanced by outflow of a corresponding volume of culture. Cells are constantly washed out with the out flowing liquid. In a steady-state the rate of cell washout equals the rate of formation of new cells in the system.

Organ Culture : The maintenance or growth of organ primordia *in vitro* without forming callus tissue.

Organogenesis : The development of organs or primordia from undifferentiated cells masses in tissue culture.

Passage : The transfer of cell or tissue from old culture medium to fresh culture medium within a definite time period. Maintenance of culture in medium for a definite period before transferring to fresh medium constitute a passage. This term is synonym with the term subculture.

Pathogen free : Free from specific pathogens based on specific tests for the designated pathogens.

Plantlet : The small regenerated plant while it is still within the culture tube or flask.

Plant Tissue Culture : The growth or maintenance of plant cells, tissues, organs or whole plants *in vitro.*

Plating Efficiency : It is a quantitative assessment of the percentage of cell colonies per total number of cell or protoplast plated at a definite density for a specific time period. It can be determined by counting cell colonies under microscope at the end of experiment and calculated by using the formula:

$$\text{Plating efficiency (PE)} = \frac{\text{No. of Colonies/plate}}{\text{No. of cell unit}} \times 100$$

Pathogen : Disease causing micro-organism.

Pathogen free : Free from specific pathogen based on specific tests for the designated pathogen.

pH : A measurement of the degree of acidity or alkalinity on a scale of 1-14.

Photoperiod : The length of time plants are exposed to light in a alternating light-dark interval

Primary Culture : A culture started from explant taken directly from plant body.

Primordium : Earliest stage in development of a plant organ.

Proliferation : Rapid multiplication of new units (cells, embryos, shoots etc).

Propagule : A small bit of plant that is being used for propagation.

Protoplast : The entire plant cell without its cellulosic cell wall.

Protoplast Fusion : Technique in which protoplasts are fused into a single cell.

Pseudodiploid : This describes the condition where the number of chromosomes in a cell is diploid but, as a result of

		chromosomal rearrangements, the karyotype is abnormal and linkage relationships may be disrupted.
Recon	:	The viable cell reconstructed by the fusion of a karyoplast with a cytoplast.
Reculture	:	Subdivision, into fresh medium.
Regeneration	:	In plant cultures, a morphogenetic response to a stimulus that results in the production of organs, embryos or whole plants.
Saturation Density	:	The maximum cell number attainable, under specified culture conditions, in a culture vessel. This term is usually expressed as the number of cells per square centimeter in a monolayer culture or the number of cells per cubic centimeter in a suspension culture.
Shoot Apical Meristem	:	Undifferentiated tissue, located within the shoot tip, generally appearing as a shiny dome–like structure distal to the youngest leaf primordium and measuring less than 0.1 mm in length when excised.
Shoot Tip (Apex) Culture	:	A structure consisting of the shoot apical meristem plus one to several primordial leaves, usually measuring form 0.10 to 1.0 mm in length; in instances where more mature leaves are included, the structure can measure up to several centimeters in length.
Somaclone	:	Plants derived from any form of cell culture involving the use of somatic plant cells.
Somaclonal Variation	:	It is genetic variability which is generated during tissue culture.
Somatic Cell Hybrid	:	The cell or plant resulting from the fusion of animal cells or plant protoplasts respectively, derived from somatic cells which differ genetically.
Somatic Cell Genetics	:	The study of genetic phenomena of somatic cells. The cells under study are most often cells grown in culture.
Somatic Embryogenesis	:	In plant culture, the process of embryo initiation and development from vegetative or nongametic cells.
Somatic Hybridization	:	The *in vitro* fusion of plant protoplast derived from somatic cells which differ genetically.
Somatic tissue	:	All the tissue that form the body of a plant.
Stock solution	:	Concentrated solution from which portion are used to make media.
Sterile	:	Free from microorganisms.
Sub culture	:	With respect to plant tissue culture, this is the process by which the tissue or explants is first subdivided, and then transferred into fresh culture medium.

Suspension Culture : A type of culture in which cells, or aggregates of cells, multiply while suspended in liquid medium.

Synkaryon : A hybrid cell which results from the fusion of the nuclei it carries.

Synchronous culture : A culture in which a majority of cells are in the same phase of the cycle or the embryos are in the same stage of development.

Synthetic seeds : Somatic embryos encapsulated in hydrated or desiccated coating, which protects the embryo from mechanical damage during handling and allows their germination like sexual seeds.

Tissue Culture : The maintenance or growth of tissues, *in vitro*, in a way that may allow differentiation and preservation of their architecture and/or function.

Totipotency : A cell characteristic in which the potential for forming all the cell types in the adult organism is retained.

Transfection : The transfer, for the purposed of genomic integration, of naked, foreign DNA into cells in culture. The traditional microbiological usage of this term implied that the DNA being transferred was derived from a virus. The definition as stated here is that which is in use to describe the general transfer of DNA irrespective of its source.

Transgenesis : In plant cell culture, the introduction and stable genomic integration of foreign DNA into a plant cell by any means, resulting in a genetic means.

Transformation : In plant cell culture, the introduction and stable genomic integration of foreign DNA into a plant cell by any means, resulting in a genetic modification. This definition is the traditional microbiological definition.

Variant : A cell line or plant exhibiting a stable phenotypic change whether genetic or epigenetic in origin.

Vector : Vehicle for transmission of genetic information from one cell or organism to another.

Virus–Free : A plant certified through specific test as being free of specified plant viruses.

Vitrification : A physical process by which a concentrated aqueous solution cooled to law or super low temperature directly solidifies into amorphous glary state withed crystallization; the term was earlier used to refer to hyper duration.

Suggested Reading

Alchanatis, V. Peleg, K. and Ziv M. (1994). Morphological control and menstruation of photo plantlets from tissue culture for automated micro propagation. *Plant All Tissue Organ. Cult.*, 36: 331–338.

Akita, M. and Takoyma S. (1993). Effects of medium surface level control on the mass propagation of photo tutors using a for ferment or culture technique. *Plant Tissue Cult.*, Let. 10: 242–248.

Al–Abta, S., Galpin, J.J. and Collin, H.A. (1979). Flavour compounds in tissue cultures of celery. *Plant Sc. Let.*, 16: 129–134.

Ali Khan, Sabaz., Hamid, Rashid., Fayyaz Chaudhary, M., Chaudhary, Zubeda. and Afroz, Amber. (2008). Rapid micropropagation of three elite Sugarcane (*Saccharum officinarum* L.) varieties by shoot tip culture. *African Journal of Biotechnology*, 7(13): 2174–2180.

Altman A. and Loberant, B. (1998). Micropropagation: Clonal plant propagation *in vitro*. *In*: Altman, A. (ed). *Agri. Biotenol. Marker Dekker*, New York. pp. 19–42.

Arencibia, A., Vazquez, R.I., Prieto, D., Tellez, P., Carmona, E.R., Coego, A., Hernandez, L., De la Riva, G.A. and Selman–Housein, G. (1997). Trans-genic sugarcane plants resistant to stem borer attack. *Mol. Breed.*, 3: 247–255.

Arencibia, A.D. and Carmona, E.R. (2006). Sugarcane *(Saccharum* spp.). *In*: Wang, K. (*Ed.*) Agrobacterium protocols. Humana Press, New York, pp. 227–235.

Barba, R. and Nickell, L.G. (1969). Nutrition and organ differentiation in tissue culture of sugarcane–a monocotylendon. *Planta*, 89: 299–302.

Barz, W. and Ellis, B.E. (1980). Plant cell culture and their biotechnological potential. *Ber. Deutsch Bot. Ges.*, 94: 1–26.

Bergmann, L. (1960). Growth and division of single cells of higher plants *in vitro*. *J. Gen. Physiol.*, 43: 841–851.

Bhansali, R.R. and Singh, K. (1984). Callus and shoot formation from leaf of sugarcane in tissue culture. *Phytomorphology*. pp. 167–170.

Bonnel, E., Demarly, Y. and Essad, S. (1983). Evolution anatomique des tissus foliares de canne a sucre (Saccharum sp.) cultives *in vitro*. *Can. J. Bot.*, 61: 830–836.

Bower, R., Elliott, A.R., Potier, B.A.M. and Birch, R.G. (1996). High–efficiency, microprojectile–mediated cotransformation of sugarcane, using visible or selectable markers. *Mol. Breed.*, 2: 239–249.

Burner, D.M. and Grisham, M.P. (1995). Induction and stability of phenotypic variation in sugarcane as affected by propagation procedure. *Crop Sci.*, 35: 875–880.

Burner, D.M. and Grisham, M.P. (1995). Induction and stability of phenotypic Variation in sugarcane as affected by propagation procedure. *Crop Sci.*, 35: 875–880.

Butterfield, M.K., Irvine, J.E., Garza, M.V. and Mirkov, T.E. (2002). Inheritance and segregation of virus and herbicide resistance transgenes in sugarcane. *Theor. Appl. Genet.*, 104: 797–803.

Chagvardieff, P., Bonnel, E. and Demarly, Y. (1981). La culture *in vitro* de tissus somatiques de canne a sucre (*Saccharum* sp.). *Agronomie Tropicale*, 36: 266–278.

Chen, W.H., Davey, M.R., Power, J.B. and Cocking, E.C. (1988). Control and maintenance of plant regeneration in sugarcane callus cultures. *J. Exp. Bot.*, 39: 251–261.

Croft, B.J., Magarey, R.C., Allsopp, P.G., Cox, M.C., Willcox, T.G., Milford, B.J. and Wallis, E.S. (2008). Sugarcane smut in Queensland: arrival and emergency response. *Aust. Plant Path.*, 37: 26–34.

Da Silva, C.M., Mangolin, C.A., Mott, A.S. and Machado, M.F.P.S. (2008). Genetic diversity associated with *in vitro* and conventional bud propagation of *Saccharum* varieties using RAPD analysis. *Plant Breeding*, 127: 160–165.

De Roppe, R.S. (1955). The growth and behavior *in vitro* of isolated plant cells. *Proc. Roy. Soc. Lond. B.*, 144: 86–93.

Enriquez–Obregon, G.A., Vazquez–Padron, R.I., Prieto–Samsonov, D.L., De la Riva, G.A. and Selman–Housein, G. (1998). Herbicide–resistant sugarcane (*Saccharum officinarum* L.) plants by Agrobacteriummediated transformation. *Planta*, 206: 20–27.

Fitch, M.M.M. and Moore, P.H. (1993). Long–term culture of embryogenic sugarcane callus. *Plant Cell Tiss. Org. Cult.*, 32: 335–343.

Fonnesbech, A. and Fonnesbech, M. (1980). *In vitro* propagation of *Monstera deliciosa. Hort. Science.*, 15: 740–741.

Fukui, K. (1983). Sequential occurrence of mutations in a growing rice callus. *Theor. Appl. Genet.*, 65: 225–230.

GalloMeagher, M. and Irvine, J.E. (1996). Herbicide resistant transgenic sugarcane plants containing the bar gene. *Crop Sci.*, 36: 1367–1374.

Gambley, R.L., Bryant, J.D., Masel, N.P. and Smith, G.R. (1994). Cytokinin-enhanced regeneration of plants from microprojectile bombarded sugarcane meristematic tissue. *Aust. J. Plant Physiol.,* 21: 603–612.

Gandonou, C., Errabii, T., Abrini, J., Idaomar, M., Chibi, F. and Senhaji, N.S. (2005). Effect of genotype on callus induction and plant regeneration from leaf explants of sugarcane (*Saccharum* sp.). *Afr. J. Biotechnol.*, 4: 1250–1255.

Garcia, A.A., Kido, E.A., Meza, A.N., Souza, H.M., Pinto, L.R., Pastina, M.M., Leite, C.S., Silva, J.A., Ulian, E.C. and Figuiera, A. (2006). Development of sugarcane *Saccharum spp.* Commercial cross, based on maximum likelihood approach for estimation of linkage and linkage phases. *Theor. Appl. Gent.*, 112: 298–314.

Garcia, R., Cidade, D., Castellar, A., Lips, A., Magioli, C., Callado, C. and Mansur, E. (2007). *In vitro* morphogenesis patterns from shoot apices of sugar cane are determined by light and type of growth regulator. *Plant Cell Tiss. Org. Cult.*, 90: 181–190.

Geetha, S., Padmanabhan, W., Manuel, W.W. and Ayyamperumal, A. (2000). *In vitro* production of sugarcane plants. *Sugar Tech.*, 2(3): 47–48.

Gurel, S., Gurel, E., Kaur, R., Wong, J., Meng, L., Tan, H.Q. and Lemaux, P.G. (2009). Efficient, reproducible Agrobacterium–mediated transformation of sorghum using heat treatment of immature embryos. *Plant Cell Rep.*, 28: 429–444.

Hamerli, D. and Birch, R.G. (2010). Transgenic expression of trehalulose synthase results in high concentrations of the sucrose isomer trehalulose in mature stems of field–grown sugarcane. *Plant Biotechnol. J.,* (In press).

Heinz, D.J. and Mee, G.W.P. (1968). Tissue culture differentiation and regeneration of plants in *Saccharrum* spp. *Argon. Abstr.,* p. 10.

Heinz, D.J. and Mee, G.W.P. (1969). Plant differentiation from callus tissue of Saccharum species. *Crop Sci.,* 9: 346–348.

Hey, J.W., Bischoff, K.P., Milligan, S.B. and Gravois, K.A. (2003). Efffect of tissue culture explant source on sugarcane yield components. *Euphytica.*, 129: 237–240.

Ho, W.J. and Vasil, I.K. (1983). Somatic embryogenesis in sugarcane (*Saccharum officinarum* L.) I. The morphology and physiology of callus formation and the ontogeny of somatic embryos. *Protoplasma,* 118: 169–180.

Hoy, J.W., Bischoff, K.P., Milligan, S.B. and Gravois, K.A. (2003). Effect of tissue culture explant source on sugarcane yield compounds. *Euphytica,* 129: 237–240.

Hunter, C.S. and Kilby, N.J. (1999). Their accumulation and release *in vitro* *In*: Hall, R.D. (*Ed.*) Methods in molecular biology Vol. III Plant Cell Culture Protocols. Humana Press Inc. Totowa. N.J. pp. 403–410.

Irvine, J.E., Benda, G.T.A., Legendre, B.L. and Machado, G.R. Jr. (1991). The frequency of marker changes in sugarcane plants regenerated from callus culture, II. Evidence for vegetative and genetic transmission, epigenetic effects and chemical disruption. *Plant Cell Tissue Org. Cult.*, 26: 115–125.

Iskandar, H.M., Simpson, R.S., Casu, R.E., Bonnett, G.D. and Maclean, D.J. Manners, J.M. (2004). Comparison of reference genes for quantitative real–time polymerase chain reaction analysis of gene expression. *Plant Mol. Biol. Rep.*, 22: 325–337.

Jalaja, N.C., Sreenivasan, T.V., Pawar, S.M., Bhoi, P.G. and Garker, R.M. (2006). A new sugarcane variety through somaclonal variation. *Sugartech.*, 8(2–3): 132–136

Jalaja, N.C., Sreenivasan, T.V., Pawar, S.M., Bhoi, P.G. and Garker, R.M. (2006). A New Sugarcane variety through somaclonal variation. *Sugar Tect.*, 8(283): 132–136.

James, G. (2004). Sugarcane. Blackwell Science Ltd, Oxford.

Joyce, P., Kuwahata, M., Turner, N. and Lakshmanan, P. (2010). Selection system and co–cultivation medium are important determinants of Agrobacterium–mediated transformation of sugarcane. *Plant Cell Rep.,* 29: 173–183.

Kaeppler, S.M., Kaeppler, H.F. and Rhee, Y. (2000). Epigenetic aspects of somaclonal variation in plants. *Plant Mol. Biol.*, 43: 179–188.

Kresovich, S., McGee, R.E. and Wadsworth, S.J. (1985). Comparisons of *in vitro* responsiveness of callus cultures derived from immature inflorescence and leaf base tissues of two interspecific hybrids of sugar cane (*Saccharum* spp.). *Sugar Cane,* 6: 8–10.

Krishnamurthi, M. and Tlaskal, J. (1974). Fiji disease resistant *Saccharum officinarum* varieties. Pindar subclines from tissue culture. *Proc. Int. Soc. Sugarcane Technologists.*, 15: 130–137.

Krishnamurthy. (1994). Alteration of conventional and unconventional means of breeding in sugarcane. Proc. SABRAO, XVII, Int. Congr. and WSAA Symp. Held at Academia Sinica, Nankang, Taipei, Taiwan, Nov. 16–19, 1(35): 285–289.

Lal, M., Singh, R.K., Srivastava, Shraddha, Singh, Neetu, Singh, S.P. and Sharma, M.L. (2008). RAPD marker based analysis of micropropogated plantlets of Sugarcane for early evaluation of genetic fidelity. *Sugar Tech.*, 10: 99–103.

Lal, N. (2003). High frequency of plant regeneration from sugarcane callus. *Sugar Tech.* 5(1–2): 89–91.

Larkin, P.J. and Scowcroft, W.R. (1981). Somaclonai variation: a novel source of variability from cell cultures for plant improvement. *Theor. Appl. Genet.,* 60: 197–214.

Liu, M.C. (1983). Sugarcane. *In*: Sharp, W.R., Evans, D.A., Ammirato, P.V. and Yamada, Y. (*Eds*) Handbook of plant cell culture, Vol. 2, crop species. Macmillan, New York, pp. 572–605.

Liu, M.C. and Chen, W. (1976). Tissue and cell cultures as aids to sugarcane breeding I. Creation of genetic variability through callus cultures. *Euphytica.,* 25: 393–403.

Luehrsen, K.R. and Walbot, V. (1993). Firefly luciferase as a reporter for plant gene expression studies. Promega Notes, 44: 24–29.

Mannan, S.K.A. and Amin, M.N. (1999). Callus and shoot formation from leaf sheath explants of sugarcane (*Saccharum officinarum*) *in vitro. Indian Sugar*. pp. 187–192.

Mudge, S.R. and Birch, R.G. (1998). T–DNA tagging and characterisation of a novel meristem–specific promoter from tobacco. *Aust. J. Plant Physiol.,* 25: 637–643.

Murashige, T. and Skoog, F. (1962). A revised medium for rapid growth and bioassays with tobacco tissue cultures. *Physiol Plant,* 15: 473–497.

Murashige, T. and Skoog, F. (1962). A revised medium for rapid growth and bioassay with tobacco tissue cultures. *Physiol. Plant*., 15: 473–497.

Nadar, H.M., Soerpratopo, S., Heinz D.J. and Ladd, S.L. (1978). Fine structure of sugarcane (*Saccharum* sp.) callus and the role of auxin in embryogenesis. *Crop Science,* 18: 210–216.

Nagai, C., Ahloowalia, B.S. and Tew, T.L. (1991). Somaclonal variants from an intergeneric hybrid: *Saccharum* spp. Erianthus arundinaceum. *Euphytica,* 53: 193–199.

Naik, G.R. and Babu, K.H. (1989). Redifferentiation of NaCl tolerant sugarcane plant from callus derived resistant lines. *Current Science*., 57: 432–434.

Pammeter N.W. and Allison J.C.S. (2002). Effect of treatement potentially influencing the supply of assimilate on its partitioning in sugarcane. *Journal of Experiment Botany*, 53(366): 123–129.

Pawar, S.V., Patil, S.C., Jambhale, V.M., Naik, R.M. and Mehetre, S.S. (2002). Rapid multiplication of commercial sugarcane varieties through tissue culture. *Indian Sugar*. pp. 183–186.

Pawar, S.V., Patil, S.C., Jambhale, V.M., Naik, R.M. and Mehetre, S.S. (2002). Rapid multiplication of commercial sugarcane varieties through tissue culture, Indian Sugar, June, 2002, pp. 183–186.

Petrasovits, L.A., Purnell, M.P., Nielsen, L.K. and Brumbley, S.M. (2007). Production of polyhydroxybutyrate in sugarcane. *Plant Biotechnol J.,* 5: 162–172.

Raghu, S., Jayaram, S., Ramkumar, S., Prabakaran, P. and Vekatesalu, V. (2006). Influence of spacing on growth and yield of sugarcane raised through *in vitro* micropropagation. *Sugar Tech.,* 8(1): 82–84.

Raghuwanshi, A. and Birch, R.G. (2010). Genetic transformation of sweet sorghum. *Plant Cell Rep.*, 29: 997–1005.

Ramanand, and Lal, M. (2004). An efficient protocol for *in vitro* micropropagation of sugarcane. *Sugar Tech.*, 6(1–2): 85–87.

Ramanand, Kureel, N., Subhanand, N., Lal, M. and Singh, S.B. (2006). Plant regeneration through leaf callus culture in sugarcane. *Sugar Tech.*, 8(1): 85–87.

Ramgareeb, S., Snyman, S.J., Van Antwerpen, T. and Rutherford, R.S. (2010). Elimination of virus and rapid propagation of disease–free sugarcane (Saccharum spp. cultivar NCo376) using apical meristem culture. *Plant Cell Tiss Org. Cult.*, 100: 175–181.

Raza, G., Ali, K., Mukhtar, Z., Mansoor, S., Arshad, M. and Asad, S. (2010). The response of sugarcane (*Saccharum officinarum* L.) genotypes to callus induction, regeneration and different concentrations of the selective agent (geneticin–418). *Afr. J. Biotech.*, 9(51): 8739–8747.

Rikiishi, K., Matsuura, T., Maekawa, M. and Takeda, K. (2008). Light control of shoot regeneration in callus cultures derived from barley (*Hordeum vulgare* L.) immature embryos. *Breed Sci.*, 58: 129–135.

Rossouw, D., Kossmann, J., Botha, F.C. and Groenewald, J.H. (2010). Reduced neutral invertase activity in the culm tissues of transgenic sugarcane plants results in a decrease in respiration and sucrose cycling and an increase in the sucrose to hexose ratio. *Funct. Plant Biol.*, 37: 22–31.

Rudus, I., Weiler, E.W. and Kepczynska, E. (2009). Do stress–related phytohormones, abscisic acid and jasmonic acid play a role in the regulation of Medicago sativa L. somatic embryogenesis? *Plant Growth Reg.,* 59: 159–169.

Sanger, R.S., Senger K., and Garg, S.K. (2011). Biotechnological approaches for high sugarcane yield. *Plant Sciences Feed*, 1(7): 101–111.

Shiromani, W., Basnayake, V., Moyle, R. and Robert G. Birch (2011). Embryogenic callus proliferation and regeneration conditions for genetic transformation of diverse sugarcane cultivars. *Plant Cell Rep.,* 30: 439–448.

Shukla, R., Khan, A.Q. and Garg, G.K. (1994). *In vitro* clonal propagation of sugarcane: Optimization of media and hardening of plants. *Indian Sugar,* 44: 113–116.

Singh, A.K. (2005). Physiological and Biochemical basis of *in vitro* morphogenesis in sugarcane hybrids. *Ph.D. Thesis*, MJP Rohilkhand Univ., Bareilly.

Singh, A.K. (2005). Physiological and biochemical basis of *in–vitro* morphogenesis in sugarcane hybrid. Ph.d. Thesis. MJP Rohilkhand University, Bareilly.

Singh, S.B. and Lal, M., Ramanand and Shubhanand, N. (2006). Raising breeder seed nursery of sugarcane through tissue culture technique. All India

Seminar on Efficient practices in Sugarcane, STAI, 25 March 2006, UPCSR, Shahjahanpur, pp. 113–118.

Snyman, S.J., Meyer, G.M., Richards, J.M., Haricharan, N., Ramgareeb, S. and Huckett, B.I. (2006). Refining the application of direct embryogenesis in sugarcane: effect of the developmental phase of leaf disc explants and the timing of DNA transfer on transformation efficiency. *Plant Cell Rep.,* 25: 1016–1023.

Sugiharto, B., Ermawati, N., Mori, H., Aoki, K., Yonekura–Sakakibara, K., Yamaya, T., Sugiyama, T. and Sakakibara, H. (2002). Identification and characterization of a gene encoding drought–inducible protein localizing in the bundle sheath cell of sugarcane. *Plant Cell Physiol.,* 43: 350–354.

Tarique, H.M., Mannan, M.A., Bhuiyan, M.S.R. and Rahaman, M.M. (2010). Micropropagation of sugarcane through leaf sheath culture. *Int. J. Sustain. Crop Prod.,* 3(2): 13–15.

Tawar, P.N., Sawant, R.A., Dalvi, S.G., Nikam, A.A., Kawar, P.G. and Deverumat, R.M. (2008). An assessment of somaclonal variation in micropropagated plant of sugarcane by RAPD markers. *Sugar Tech.,* 10(2): 124–127.

Taylor, P.W.J., Geijskes, J.R., Ko, H.L., Fraser, T.A., Henry, R.J. and Birch, R.J. (1995). Sensitivity of random amplified polymorphic DNA analysis to detect genetic variation in sugarcane during tissue culture. *Theor. Appl. Genet.,* 90: 1169–1173.

Taylor, P.W.J., Ko, H.L., Adkins, S.W., Rathus, C. and Birch, R.G., (1992). Establishment of mebryogenic callus and high protoplast yielding suspension cultures of sugarcane (*Saccharum* spp. hybrids). *Plant Cell Tissue Organ Culture,* 28: 69–78.

Taylor, P.W.J., Geijskes, J.R., Ko, H.L., Fraser, T.A., Henry, R.J. and Bich, R.G. (1995). Sensivity of random amplified polymorphic DNA analysis to detect genetic change in sugarcane during tissue culture. *Theor. Appl. Genet.,* 90: 1169–1173.

Taylor, P.W.J. and Duke, S. (1993). Developement of an *in vitro* culture technique for conservation of *Saccharum spp. Plant Cell, Tissue and Organ Culture,* 34: 217–222.

Tisserat, B. (1981). Date Palm tissue culture, U.S. Deptt. of Agric. Agricultural Research Service, Oakland, California.

Vickers, J.E., Grof, C.P.L., Bonnett, G.D., Jackson, P.A. and Morgan, T.E. (2005). Effects of tissue culture, biolistic transformation, and introduction of PPO and SPS gene constructs on performance of sugarcane clones in the field. *Aust. J. Agric. Res.,* 56: 57–68.

Visessuwan, R., Chiemsombat, P., Naritoom, K. and Surija Chaijakorn, M. (1999). Role of growth regulators in meristem culture and production of virus free sugarcane germplasm. *Sugar Tech.,* 1(3): 82–88.

Von Arnold, S.V. and Erikson, T. (1981). *In vitro* studies of adventitious shoot formation in Pinus contorta. *Can. J. Bot.,* 59: 870–874.

Wagih, M.E. and Adkins, S.W. (1998). Embryogenic callus culture for Fiji disease virus elimination in sugarcane. *4th Breeding and Germplasm Workshop, ISSCT, Mauritius, 22nd* May, pp. 1–15.

Wagih, M.E., Ala, A. and Musa. Y. (2004). Regeneration and evaluation of sugarcane somaclonal variants for drought tolerance. *Sugar Tech.,* 6(1–2): 35–40.

Wagih, M.E., Ala, A. and Musa. Y. (2004). Regeneration and evaluation of sugarcane somaclonal variants for draught tolerance. *Sugar tech.,* 6(102): 35–40.

Wang, M.L., Goldstein, C., Su, W., Moore, P.H. and Albert, H.H. (2005). Production of biologically active GM–CSF in sugarcane: a secure biofactory. *Transgenic Res.*, 14: 167–178.

White, P.R. (1963). In the cultivation of animal and plant cells (2nd edition) New York Ronald Press.

Wu, L. and Birch, R.G. (2007). Doubled sugar content in sugarcane plants modified to produce a sucrose isomer. *Plant Biotechnol. J.,* 5: 109–117.

Yao, K. Chen, H. and Mc Mallen C.R. (1991). Reservation of dine shoots in cultured authors of intermediate wheat grass (Agropyron inter medium (Host) Beav) *Plant All Tissue Organ Cult.* 26: 189–193.

Yadav, S., Saini, N. and Jain, R.K. (2004). Low cost multiplication and RAPD analysis of micropropagated plants in sugarcane. *Physiol. Mol. Boil. Plants,* 10(2): 269–276.

Zhang, L, Xu, J. and Birch, R.G. (1999). Engineered detoxification confers resistance against a pathogenic bacterium. *Nat. Biotechnol.*, 17: 1021–1024.

Zhang, S.Z., Yang, B.P., Feng, C.L., Chen, R.K., Luo, J.P., Cai, W.W. and Liu, F.H. (2006). Expression of the Grifola frondosa trehalose synthase gene and improvement of drought–tolerance in sugarcane (*Saccharum officinarum* L.). *J. Inter. Plant Biol.*, 48: 453–459.

Subject Index

D

E

F

G

N

O

P

Q

R